Should Global Agriculture be Liberalized?
Theories, models and realities

Should Global Agriculture be Liberalized?
Theories, models and realities

Jean-Marc Boussard
Senior Researcher, INRA
(French National Agronomic Research Institute)
Paris, Fellow of the French Academy of Agriculture

Françoise Gérard
Researcher, CIRAD
(International Research Center for International Developement)
Paris

Marie-Gabrielle Piketty
Researcher, CIRAD
(International Research Center for International Development)
Sao Paulo

Science Publishers

Enfield (NH)　　　Jersey　　　Plymouth

Science Publishers
234 May Street
Post Office Box 699
Enfield, New Hampshire 03748
United States of America

www.scipub.net

General enquiries : *info@scipub.net*
Editorial enquiries : *editor@scipub.net*
Sales enquiries   : *sales@scipub.net*

Published by Science Publishers, Enfield, NH, USA
An imprint of Edenbridge Ltd., British Channel Islands
Printed in India

© 2008 reserved

ISBN 978-1-57808-542-2 (PB)

Library of Congress Cataloging-in-Publication Data

Should global agriculture be liberalized?: theories, models and realities/ editors Jean-Marc Boussard, Françoise Gérard and Marie-Gabrielle Piketty. -- 1st ed.
   p.cm.
  Includes bibliographical references and index.
  ISBN 978-1-57808-542-2 (pbk.)
 1. Agriculture--Economic aspects. 2. Free trade. 3. International trade. I. Boussard, Jean-Marc. II. Gérard, Françoise, Dr. III. Piketty, Marie-Gabrielle.
   HD1415.S52 2008
   338.1--dc22
                                                                    2008029172

Published by arrangement with Editions Quae, Paris
French edition: © Cirad, Paris, 2005
         ISBN 2-87614-619-3

Update (Chapter 8) added for the English edition in 2008

All rights reserved. No part of this publication may be reproduced, stored in a retrieval system, or transmitted in any form or by any means, electronic, mechanical, photocopying or otherwise, without the prior permission of the publisher, in writing. The exception to this is when a reasonable part of the text is quoted for purpose of book review, abstracting etc.

This book is sold subject to the condition that it shall not, by way of trade or otherwise be lent, re-sold, hired out, or otherwise circulated without the publisher's prior consent in any form of binding or cover other than that in which it is published and without a similar condition including this condition being imposed on the subsequent purchaser.

Acknowledgements

This small book presents the synthesis of team work by CIRAD[1] and INRA[2] researchers. Apart from the authors, the team comprised:
- Ane Catherine Christensen, great expert in GTAP databases;
- Mourad Ayouz, statistician, economist and talented programmer;
- Tancrède Voituriez, subtle researcher, statistician and also expert in GTAP databases;
- Abigaïl Fallot, author of many recommendations for the use of our models in negotiations on the greenhouse effect;
- Cirad trainees participating episodically but always in a useful manner;

Without them, the ID^3 model that plays the central role in this presentation would not have been constructed.

It would not have been constructed either without the financial assistance of the following:

---

[1]International Center of Agricultural Research for Development
[2](French) National Agronomic Research Institute

- the GICC Program (Management and Impacts of Climate Change) that enabled us to work on a pre-model;
- the Pluriagri Association and the Ministry of Agriculture that provided assistance in turns;

We benefited also from valuable comments from the Cirad Review committee - particularly, Benoît Daviron, and various personalities such as Pierre Leroy, Bruno Vindel and Jean-Christophe Debard. Lastly, we should express gratitude to the late Isabel Boussard who proofread the manuscript, and Valérie Hourmant, tireless author of thousands of graphs.

The opinions reported herein are those of the authors who are solely responsible thereof.

Contents

Acknowledgements	*v*
Introduction	*xi*
I. What can be expected from the Liberalization of Agricultural Trade?	**1**
Exploiting Natural Advantages	2
Box 1: Ricardo's Parable	3
Better use of Production Factors	6
The Minimal Cost in a State of Technology	7
Price Stabilization	8
Box 2: Risk Aversion	11
Box 3: The Law of Large Numbers and Market Stabilization	13
II. Theoretical Criticism of Agricultural Liberalism	**17**
Box 4: Galiani	18
A Radical Criticism: Liberalism and Justice	18
Another Radical Criticism: Markets and Needs	21
An Apparently Incidental Criticism: do Markets Function?	22

Expectation Errors and Endogenous Price Fluctuations	25
Box 5: The Notion of Demand Elasticity	25
Box 6: Algebraic Expressions of the Cobweb Theorem	29
The Traditional Cobweb	29
The "Risk Cobweb"	29

III. The Test of Facts   35

Brief History of Liberalism	35
Price Fluctuations and Growth	40
Price Fluctuations and Production	41
Are Price Fluctuations Endogenous?	45
Are Price Fluctuations Caused by Climate?	46

IV. Designing an Economic Model for International Trade   53

Production	55
Box 7: The CES Function	57
Box 8: "First-order Conditions"	59
Consumer Demand	59
Box 9: The LES Function and Consumer Behavior	60
Income Determination	61
International Trade	64
Data Sources	65
Box 10: The GTAP Database	66
Dynamics: The Role of Time in the Model	70
The Standard Model	76

V. How can Theory and History be Introduced in a Standard Model?   79

Expectation Errors	80
Capital Accumulation	85
Box 11: The Markowitz Model	86
The Role of Firms, Banks and Financial Markets	89
Illustration of Differences	91

	ix

VI. A Choice of Results — 97
   Does the Detailed Model Confirm the Results of the Three-Region Model? — 98
   Why may Europe Benefit from Liberalization? — 100
   Effects of Liberalization in Sub-Saharan Africa — 106
   To Conclude — 112

VII. Could We Do Better? — 117
   The Futures Market Option — 120
   Box 12: The Algebra of Futures Markets — 124
   Stocking and Destocking — 126
   Production Quotas — 130
   Are Futures Markets and Quotas Different in their Principles? — 135
   To Conclude — 138

VIII. Recent Developments — 139

Conclusion — 149

References — 153

Introduction

Quos vult perdere, Jupiter dementat. "Those He wishes to ruin, Jupiter blinds them"

The world is worried about its agriculture. In developed countries, a pervasive excess production leads to wild competition for external markets. No care is taken to eschew the ridiculous situation of having to spend two dollars to get one euro or two euros to get one dollar, while the money dealer round the corner does it at one to one. In developing countries, production is insufficient and hunger remains a major scourge. Yet, insufficient production is not the issue here because countries that are unable to feed their population are complaining at the same time of not being able to export their agricultural commodities to rich countries.

Many people, very wise and full of goodwill, are aware of the absurdity of this situation. They constitute some sort of world intelligentsia seeking to find solutions to these too obvious weaknesses of current political organization. It is hoped that this book will suggest a few and somewhat unprecedented solutions.

Liberalizing trade in agricultural commodities and allowing the market to manage agricultural production like every other economic activity is the solution most often envisaged nowadays. For a long time, in the vast majority of countries in the world, agriculture has been more or less excluded from the market, to the extent that one may wonder whether a return to "natural law" would not resolve many problems. Besides, all students in economics are well aware that the "invisible hand[1]" is both strong and efficient. It ensures that there would be no wastage, as those that we witness when stocks of butter and meat flour have to be disposed of. It seems to guarantee that the consumer would always pay the "fair price" thanks to competition that prevents whosoever from earning any unjustified rent. Based on this reasoning, we may wonder how we found ourselves in this situation, depriving ourselves of the benefits of the market, a natural institution, which emerges spontaneously without any need for particular effort. Yes, the exclusion of agriculture from the market must have been a matter of will. Where did such will stem from?

There are two opposing view points in response to this question. For some, the exclusion of agriculture from the market was decided by malice, because it benefited various social categories, top among whom were obviously the farmers. But farmers were not the only beneficiaries and perhaps not even the major ones. Candidates on the infamous list also include landowners, agricultural suppliers, the chemical industry and bureaucrats of all sorts, who always benefit from settlements and administrative complications. It was in the interest of these people to extort a maximum of money from taxpayers, under false pretence, and to share the spoils even at the cost of flagrant inefficiency. This idea is indeed the most widespread among authorities of all kinds. Indeed, it justifies the return to the sound discipline of a real market from which we wrongly excluded ourselves.

For the others, by contrast, the market was discarded

---

[1] A term used by Adam Smith (1723-1790) to illustrate automatic regulation brought about by free markets to the economic system.

because it was not functioning satisfactorily, in any case, not as it is expected to function. Nowadays, such an idea is profoundly iconoclastic. Did the fall of the Berlin Wall not consolidate the triumph of market economies over communist socialism which collapsed into bureaucracy for having unwisely proclaimed its ambition to free itself from this law of nature? And is not the prosperity of Anglo-Saxon countries, which everyone can see, due to the profound liberalism we observe therein?

Before responding positively to these questions, there is an observation that needs to be made: the effective liberalism of Anglo-Saxon countries has nothing to do with unrestrained laisser-faire. The societies in question are under extreme control. Many precautions are taken to ensure that individual initiative, without being unnecessarily restricted, is carried out to the benefit of society without undermining it. This is true in general and particularly in the area of agriculture and food economy, sectors where the market actually plays a minimal role in determining supply and producers' remuneration. Paradoxically, in the 60s, the agricultural market played a much greater role in the Soviet Union, where "kolkhozian markets" accounted for a major part of the supply and farm income sources, than in the United States where almost all prices were controlled by Government and severe limits to supply increases were imposed. Since sixty years, American agricultural policies have been imitated in this domain by all developed countries and by many developing countries.

The question then is to know whether the prosperity of the currently "rich" countries was achieved "thanks to" or "in spite of" the characteristics of these modern agricultural policies. Such is the subject of this book. There are several ways of tackling this kind of question and we will endeavor to do so from different angles.

In chapter I, the focus is on consideration of "liberal" arguments: why should we wish to liberalize agriculture now and what do we expect from such a change in world policies? Here, there are situational and anecdotal reasons such as the existence of huge excesses or of gigantic budgetary costs. There are also bureaucratic reasons. There is an international institution in charge of liberalizing everything and it has almost nothing left to liberalize apart from agriculture. However, apart from such superficial considerations, one needs to tackle the question of why at present, for the vast majority of economists, "liberalization of agriculture would be a good thing". The fundamental reason is that many economic models indicate so. Consequently, we need to find out why and how these economic models have been constructed. Long sections of the book are devoted to this. Although a little arid, they are necessary to apprehend the problem. Before, we would start by considering what opponents of liberalism say.

Indeed, there are many authors who hold a contrary view to the preceding assertions. We have considered their arguments in chapter II. Chapter III provides some historical data to back them. We need to recall here the origin of "Roosevelt-styled" policies which precisely were based on the idea that conditions were not conducive for tapping the beneficial effects of liberalization. By analyzing a little deeply the ideas of theoreticians of that period and interpreting them in the light of modern mathematical knowledge, we see that such situation may be confirmed by reality.

Such consideration of facts is the focus of chapter III. It is based on an analysis of available historical data. They show that food consumption is largely independent of the prices of staple foodstuff and, especially, that the production of this foodstuff depends, to a limited extent, on price levels, but also greatly on their volatility. These are very unfavorable conditions for the functioning of markets. We will study the consequences attendant thereto.

However, such partial analyses do not have the convincing power of fabulous figures provided by international agencies based on econometric models that bring into play dozens of thousands of equations. To consider the rationale of these calculations, we have to examine the manner in which these models are constructed and, consequently, the importance and impact of their results. The somewhat arduous and abstract exercise of demystifying them, which consists in demonstrating how do they work by "unveiling their contents", is the focus of chapter IV, which the hurried reader may omit.

We see from an analysis of the "standard" model that liberalization would be an excellent thing if markets functioned appropriately and if information communicated to producers and consumers through prices were reliable. However, this is not possible, and models have to be modified to take that into account.

By revisiting the analyses in chapters II and III and applying them to the model described in chapter IV, chapter V describes the (apparently minor) modifications, which have to be effected on the standard model. Yet, with the same data and the same kind of "marginally" modified model, very different results from those that constitute the basis of the optimistic conclusions complacently provided by international institutions to back liberalization programs are obtained.

In chapter VI, these results are detailed out, broken down into regions, and especially between "the rich" and "the poor". Indeed, the arguments evoked hitherto, bear only on national incomes and the general "well-being". However, in a matter such as the agricultural liberalization, it is difficult to obfuscate the fact that a change in food prices has a considerable bearing on the distribution of income. This is one of the arguments raised to back liberalization policies: such policies should help to combat poverty by providing cheap food. However, we will see that such hypothesis itself is debatable.

Lastly, chapter VII tackles the issue of possible options.

It begins by addressing the issue of whether there are liberal solutions to "market failures" that can be pointed out in the economic policies proposed by the WTO school. Could crop insurance and "futures markets" nullify the disadvantages of liberalization while preserving its indisputable benefits? Similarly, could progressive income taxes re-establish some minimum equity in the system? We will see that this remains problematic.

Well, if liberalization has limited advantages and many disadvantages and if current policies are too costly and a source of wastage, is there any other solution? What is the cause of the general dissatisfaction concerning agricultural policies? In order to understand this, we need to use the theory of production to demonstrate that in agriculture (and in agriculture alone!), there are no reasonable limits to production increases with guaranteed prices and without restriction of quantity. This is why the only reasonable option other than price fluctuations seems to be "supply control" – put simply, a general system of production quotas, which can very well accompany a non-negligible role assigned to markets for "marginal" adjustments. We shall conclude that the relations between "production quotas" generally abhorred by "liberals" and futures markets often presented as universal remedies to market failures, may be narrower than they appear to be.

Such panoramic view of agricultural trade obfuscates an increasingly important perspective, namely the "multifunctionality" of agriculture. Agriculture, everywhere in the world, covers vast expanses. By the same token, it creates "sub-products" obtained without reference to the markets, amenities and nuisances which, clearly, affect the general well-being. This is one argument often used by "anti-liberals" to refuse to submit agriculture to "market laws". The argument is serious. Many fanatics of liberalism are so conscious of it that they are seeking and proclaiming to have found methods of creating some kind of virtual markets where it will be possible to manage amenities and nuisances in a decentralized manner. We wanted to ignore this issue to avoid digressing and confusing the reader with considerations that would oblige us to transcend the limited scope of traditional economic analysis. However, it is also of great importance. The models we have constructed also make it possible to determine some of the consequences of greenhouse effects.

Current global agricultural policies, far from building a harmonious world, are very likely to revive the old curse described by Robert Malthus: the limits to food supply stands as an ultimate barrier to economic growth. This barrier will not function in a gentle manner, allowing politicians time to reflect (a time which, besides, is very often put to bad use). It will close suddenly and dramatically, prompted by an incident such as a drought or flood, something that under "normal" circumstances would not have sufficed to provoke a real shortage. It would become an absolute disaster because of the sole fact that the disaster was only waiting to be triggered[1]. In such circumstance, common sense requires that we distance ourselves from the situation that transforms a common natural phenomenon into a disaster. It would therefore be absurd to renounce traditional agricultural policy instruments to subject oneself without guarantee to "blind market forces", though we must be aware of the eminently perfectible nature of current policies.

---

[1] Exactly like a stock market crash, often "triggered" by fraud or the bankruptcy of an imprudent company. However, the proportion of dishonest brokers and adventurous industry managers is no doubt constant in time. If they trigger a crash at a given place at a given moment, it is because circumstances so permit. In this respect, we may review the remarks made by Daniel Zajdenweber (2001) concerning the Lisbon earthquake in 1755 and the controversy between Voltaire and Rousseau on the "wickedness of nature": you wouldn't say it is nature's fault, said Rousseau, if hundreds of thousands of people crowded together in wooden houses at the bottom of a narrow valley, would you?

CHAPTER I

What can be Expected from the Liberalization of Agricultural Trade?

Trade is comparable to technical progress

The idea that any obstacle to free trade, particularly in agriculture, would lead to negative consequences is not recent. It can be traced back at least to the 18th century, a period when the awkward and finicky royal bureaucracy seemed to take a perverse pleasure in complicating the task of traders. In France, grain could not be moved from one province to another without the prior authorization of the District administrator (the King's Local Representative), who delayed his decisions and sometimes only gave authorization against cash...The situation was similar throughout Europe. It provoked frustrations and sarcasms, theorized by the "economists" – a very active school of thought, led by very influential persons like François Quesnay (1694-1774), Louis XV's personal doctor who, besides, propounded the basic concepts of national accounting.

The disciples of Quesnay were later referred to as "physiocrats" – people who wanted to govern "scientifically", based on the principles of social physiology. They undoubtedly made their mark on the history of economic thought. In particular, they laid the foundation of liberalism through Adam Smith (1723-1790) who read and discussed them. However, the real theory of international trade, on which current practice is based, was developed by David Ricardo. He is the author of the most powerful justification of trade ever provided by the economists.

Exploiting Natural Advantages

David Ricardo (1772-1823) was a Portuguese banker who emigrated to London in the late 18th century. His theory, which is both marvelously simple and profound, is entirely contained in a parable: in England as in Portugal, fabric or wine can be produced. However, it is "relatively" (or "comparatively", which explains the origin of the expression "comparative advantage" used to describe this notion) easier to produce wine in Portugal than in England. By the same token, the total quantity of fabric and wine that can be produced in the two countries put together would be greater if the Portuguese specialized in wine production and the English in fabric production (see Box 1), even if one of the two countries is more efficient than the other in the production of both products, simply by virtue of the fact that it is different from the other.

In reality, Ricardo's message is that profit in trade does not depend on prices, which concern only the physical quantities of goods and not values. In this way, trade is comparable to technical progress, which does not depend upon prices either. Thus, if the English produce greater quantities of fabric than they need for local consumption in order to exchange the surplus for wine, it simply means a more efficient manner of securing wine than having to produce it at home. Technical progress cannot be "bad". Trade can only be "good".

Such reasoning constitutes the basis of the action of the World Trade Organization (WTO) towards trade liberalization. It shows that contrary to a very frequent argument, trade can only lead to a win-win situation (see Box 1). Obviously, some may gain less than others, depending on the prices adopted. However, this must not hide the fact that if there is generally a winner, there can normally be no loser.

> **BOX 1   RICARDO'S PARABLE**
>
> David Ricardo (1817) explains that "this same rule for determining the relative value of products within a country does not apply to the relative value of products traded between two or more countries... If Portugal had no trade links with other countries, instead of using most of its capital and resources to produce wine, then to sell it and buy fabric and various products from outside, it would be obliged to use part of its capital and resources to produce such goods, with probably lower quality and quantities...".
>
> "Let us assume that circumstances are such that in England, producing a certain quantity of fabric would require 100 men per year, while producing a certain quantity of wine would require 120 men per year. In Portugal, producing a certain quantity of wine would require only 80 men per year and producing fabric would require 90 men per year. It would therefore be in Portugal's interest to export wine in exchange for fabric. Such trade would take place in spite of the fact that production of the imported good could have entailed less work in Portugal than in England. Indeed, even by importing a product that requires just 90 men in Portugal to produce from a country where production requires 100 men, it is still more advantageous for the country to use its resources to produce wine to sell and obtain more fabric than it would otherwise have got by diverting part of its wine-production resources to produce the fabric."
>
> The table below presents such "production possibilities".
>
> | Country | Fabric
Unit cost in men/year | Wine
Unit cost in men/year |
> |---|---|---|
> | England | 100 | 120 |
> | Portugal | 90 | 80 |
>
> Based on these figures, let us assume again that there are 170 men in Portugal and 220 men in England and that where trade is possible, England would specialize entirely in fabric production and Portugal in wine production. We would then have the following "scenarios":
>
> | | Without trade | | With trade | |
> |---|---|---|---|---|
> | Production | Wine | Fabric | Wine | Fabric |
> | in England | 1 | 1 | 0 | 2.2 |
> | in Portugal | 1 | 1 | 2.125 | 0 |
> | Total | 2 | 2 | 2.125 | 2.2 |
>
> We may make other specialization assumptions. However, whatever the case, the quantity of goods produced "with trade" is greater than the quantity produced "without trade".

This argument departs significantly from the "carpet dealer's" point of view (what I gain by bargaining the price of a carpet that I want to buy is lost by the seller and *vice versa*), which very often underlies international negotiations and which is publicized by the media. The Ricardian approach transcends the issue of "exploitation" based on this carpet-dealer argument, which says, for instance, that "North-South" trade between rich countries of the North and poor countries of the South benefits exclusively the North because the latter is "stronger".

Trade is advantageous for both parties

Obviously, there are limits that should not be exceeded in the rates of conversion between two commodities ("terms of trade"): if in producing fabric that would be traded for wine, the English must make more efforts than would normally be required to produce directly the wine they need, then trade would be of no benefit to them. This also applies to Portugal. However, if there is quite a great difference between the two countries, then the possible terms of trade range itself (between the rate that disfavors the English and that which disfavors the Portuguese) is quite wide (Box 1). Within this range, what is important is that trade is advantageous for both parties. Whatever the case, given that nobody is obliged to trade, and that each party is free to refuse it, the less favored party cannot find itself in a position worse than would otherwise be without trade.

It can thus be affirmed that even if countries of the North impose their prices on countries of the South under conditions that are not advantageous for the latter, the people of the South would be worse off if they were completely prohibited from trading with the North.

Based on Ricardo's parable, it is also difficult to understand the argument that trade leads to a situation whereby "countries of the South export their unemployment to the North". The apologue assumes that resources of all kinds are fixed in the two countries and are used to the maximum of their possibilities. Under such conditions, the labor which would no longer be employed in the vineyards in England because of "delocalization" of wine production would be diverted to the fabric industry that would be stimulated. The same reasoning applies to the Portuguese labor laid-off from factories "destroyed by (unfair?) competition with the Manchester factories". Such labor would then be redeployed in the vineyards of Porto[1].

Of course, none of the considerations developed in the newspapers concerning "market conquest" and other strategic operations is taken up here. Ricardo undoubtedly would have considered "export subsidy" policies with commiseration. The objective of such policies can be none other than to earn foreign "currency"- foreign money. However, if one were to spend the equivalent of 2 dollars in euros to get just 1 dollar, where is the benefit? This would be tantamount to firing on one's toes like an awkward cowboy!

This kind of argument is universal: it applies to all products, and not only to Portuguese wine or English fabric. However, it applies more to agriculture than to any other product. There is nothing that *a priori* makes one think that it would be easier to produce electronic chips in Hong-Kong than in Wolverhampton. By contrast, it is clear that it is easier to grow cocoa in Côte d'Ivoire than in the Netherlands. This explains why agricultural liberalization, more than any other, appears at first sight more natural. This also explains why it may appear paradoxical that in the long rounds of negotiations that have been going on for sixty years and which have resulted in the liberalization of almost everything since the end of the Second World War, agriculture is the last important sector where the issue is still unresolved.

---

[1] More recently, this argument was vigorously taken up by Krugman and Obstfeld (2003). Naturally, it is assumed that the workers are quite flexible to switch from one profession to the other.

We will revisit this issue later. However, at this stage, it is important to note that if liberalization of industrial products finally seemed easier than that of agricultural produce, it is because arguments in favor of liberalization have at the same time slightly evolved to the extent that it is sometimes difficult to perceive the great simplicity of the story told by Ricardo.

Better use of Production Factors

Ricardo deemed it appropriate to use an example relating partially to agriculture because, as we have just noted, "advantages" connected with geographical location are obvious in this sector. We may broaden the question by seeking to know the source of the advantages that would be tapped from trade. Undoubtedly, there is an easily perceptible "natural" source, but are there any other ones? The answer here is again yes, but requires a slightly higher degree of abstraction to be understood. Unlike the other justification relating to the advantage of "the free gift of Nature", this justification of international trade particularly challenges social organization, and by extension the "Princely fiat" (Unless it is assumed that the Prince himself is a gift of Nature!).

Indeed, according to this interpretation, the advantage that some countries have in certain activities derives from the fact that their resources, particularly labor and capital, are different from those of other nations. For instance, for purely fortuitous historical reasons, Hong Kong does not have much capital but can rely on an abundant and submissive manpower, while Paris has an abundance of machinery but manpower is rare and undisciplined. Under such conditions, delocalizing to Hong Kong the manufacture of labor-intensive goods, such as computer keyboards, will allow for better use of Chinese manpower. At the same time, memory chips, which require extremely complicated machines, will be produced in Paris, to better take advantage of the high technological resources of that city.

Apparently, this new illustration is very similar to Ricardo's. In both cases, the optimal exploitation of "differences" allows for mutually beneficial productivity gains for all. Here again, trade acts like technical progress. However, the illustration presupposes that the cultural and institutional context that underpins the difference should be unchanging: there is hardly any possibility of growing the vine in Scotland[1], whereas the differences between China and France may be eliminated and the importance of trade ruined by relocating the machines from Paris to Hong Kong or by allowing workers to migrate from Hong Kong to Paris.

A common and universal language: the market.

It is probably easier to move goods than cultures, and this is no doubt a powerful reason for accepting globalization, from this stand point. In addition, this story makes it possible to understand that the problem is no longer as simple as comparing the output of the vineyards and the quality of wine produced in Portugal and England. To know whether it is more advantageous to produce in one location and not in the other, it would be necessary to compare an incredible amount of parameters, such as the quality and the technology used to manufacture computer keyboards and the electronic chips. Such exercise is perhaps within the capacity of a specialist in the domain, but surely not the general public. Therefore, one other condition necessary to validate the preceding argument is the need for a transmission medium, a common and universal language that allows for easy comparison of techniques and determination of those that may have a comparative advantage in one location and not in the other(s).

The Minimal Cost in a State of Technology

In a well-behaved market, the production cost is the "fair price".

Such a common and universal language exists: it is the market. In each market, that is, if the market is functional – and we will see that this is where the whole

---
[1]Though, after all, with genetic manipulations...

problem lies – there is a price. There are prices for all products, and "production factors", as well as for "intermediate commodities" - those goods that are used to produce other goods. Knowing the prices and production techniques, one may calculate the cost of production and see whether it is higher or lower in one location as against another. It suffices to determine where costs are lowest to automatically exploit to the maximum the advantages related to localization.

However, the minimal cost is determined through competition: if a seller displays a price higher than allowable by current technology, someone would point it out and the price would have to drop. In the final analysis, price in a well-behaved market cannot be different from "marginal cost", that is, the lowest production cost just necessary to match demand. It is difficult not to consider this price as the "fair price". Once more, whatever the remuneration of labor in Hong Kong, or of the machinery in Paris, trade is still beneficial for all if the relative price lies within a certain range outside which there would be no possible transaction.

Obviously, under such conditions, any impediment to trade can but undermine this wonderful game and usher in "distortions" which always cause either the consumer price to rise higher than technically necessary or remuneration of production factors to drop lower than acceptable. It is therefore absurd to deny that there are benefits to be gained from a free trade system combined with well-behaved markets, be it in agriculture or in any other sector.

Price Stabilization

The foregoing is even truer for agriculture than all other sectors of activity. There are no "geographical" reasons, attributable to natural conditions, to justify why computer keyboards should be manufactured in Hong Kong instead of Paris: the people of Hong Kong accept to sell their labor cheap or cannot do otherwise. However, such situation is likely to change very quickly. By contrast, growing cocoa in Cote d'Ivoire instead of Britain is quite "natural". This is true to the extent that no one has ever thought of protecting British agriculture against unfair competition with Ivorian cocoa...

In reality, in agriculture, everything that "naturally" had to be liberalized has already naturally been liberalized. What remains to be liberalized – though probably larger – is perhaps not so important from the sole point of view of comparative advantages. This explains why until now, the issue has not been paid great attention to by diplomats, already very busy liberalizing trade in manufactured products. Now that this goal has been achieved, the agenda has to be moved forward. And one begins to reflect on what is still left to be done. Price instability is however one serious reason, peculiar to agriculture, for attempting liberalization.

It is common knowledge that the prices of agricultural commodities are unstable. In Europe, no one seems to be aware of the magnitude of the problem: in developed countries at least, for the past sixty years, prices of the most important foodstuffs are stabilized through agricultural policy, in such a way that the consumer is not subject to very serious stop-go measures[1].

Figure 1 however aptly illustrates the importance of the instability of prices of agricultural commodities compared to those of industrial products.

We can see that none of these prices is constant. Car prices change depending on the technology, inflation level and consumer tastes. They however remain

---

[1] To explain such indifference, it should also be noted that food is no longer the most important expenditure for households and that agriculture is no longer the main source of food cost. This is why some economists are talking superficially about "the end of the agricultural exception". Yet, if a real shortage loomed in the horizon, it would be talked about! It is precisely stabilization policies that have warded off the "spectre of famine". This is why it may be irresponsible to abandon them.

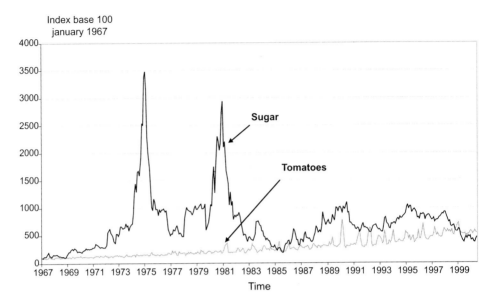

Figure 1 The agricultural exception: Comparative trends of prices of new cars in the United States, retail prices of fresh tomatoes in large American cities and sugar at the Chicago Board of Trade, from 1967 to today.

relatively stable. A car buyer in a big American city during this period, unless he was exceptionally stupid, never paid a price far different from the production cost.

This was not the case with the tomato buyer. It is impossible to admit that tomato production cost moved up from index 250 to index 800 in a few weeks, as was the case in 1989. The person who bought tomatoes at 800 in April 1989 clearly paid an undue economic rent to the producer. Conversely, the producer who sold tomatoes at 240 in January 1990 surely did not gain anything (and indeed, lost a lot of money!).

The case of sugar is quite different. Sugar price fluctuations dwarf those of tomato, which, comparatively, seem stable. These are extravagant price hikes that are not at all related to the cost of production of a purely "speculative" product.

BOX 2 RISK AVERSION

Few people would like to take a risk for risk's sake. In most cases, to encourage a person to take a risk, you must offer him a reward – glory, hope of winning great fortune or something of the sort. Economists have for long elaborated "models" of the economic man's behavior when faced with risk. The simplest and oldest of them are the works of Buffon (who was also a "naturalist", author of renown monographs on animals, and also a distinguished mathematician), in the 18th century. Buffon considers that in the face of a random profit possibility, people would maximize their profit expectation after deducting the "cost of risk". Thus, $\bar{x}$ is the profit expected from a lottery game where the standard deviation of the result is σ. As such, the decision-maker would maximize: $U = \bar{x} - A\sigma^2$, where A is the coefficient referred to as "risk aversion". The term $A\sigma^2$ represents the "cost of risk", which is a psychological cost, the risk being measured with the σ^2 variance and the price given by A, which represents the decision maker's preferences.

Coefficient A is not "dimensionless": given that σ^2 is a squared amount of money, A is the reverse of a quantity of money. There are grounds for thinking that A indeed represents something that resembles the decision maker's wealth: as a matter of fact, it is not the same thing as having a one-in-two chance of winning $10 000 if one has nothing or if one already has a great fortune. Thus, in 1780, Daniel Bernoulli, drawing inspiration from Buffon's idea, demonstrated that a beggar who finds by chance a lottery ticket with a one-in-two chance of winning $10 000 the next day would be well-advised to accept the proposal of the billionaire who offers him $4000 on the spot. Conversely, the billionaire who can afford to lose this amount is doing good business by buying at that price a five-thousand dollars profit expectation. Thus, the transaction is mutually beneficial, as is trade according to Ricardo. Insurance business and "derivatives markets" are based on this principle.

However, in ordinary life and for a producer, such risk aversion leads to regrettable consequences, particularly the need to reduce production. Consider, for instance, an entrepreneur producing quantity q at cost $C(q)$ and selling at price p. In the absence of risk, maximization of q against $pq - C(q)$ results in $p = C'(q)$: the price is equal to the marginal cost. In the face of a price risk, with a random price, of expectation $\bar{p}$ and standard deviation σ, the preceding reasoning calls for determination of q which maximizes:

$$U = \bar{p}q - A\sigma^2 q^2 - C(q).$$

This is tantamount to looking for the value of q which satisfies $\bar{p} - 2A\sigma^2 q = C'(q)$. Everything is therefore done as if there has been a drop of the price used in the economic calculation of the entrepreneur's q quantity. Naturally, this causes him to reduce his production. In addition, production very quickly reaches a production ceiling, because once this becomes substantial, the first element of the equation becomes negative.

The quantity $\pi = \bar{p} - A\sigma^2 q$ plays exactly the same role as an "adjusted" price for the associated risk. Everything is done as if the entrepreneur was basing his economic calculations on this "certain" price instead of using the "uncertain" price π. As a result, it is said that p is the "certain equivalent" of $\bar{p}$, or at least the measure thereof.

Thus, one can see that price fluctuations lead to significant gains or losses, both for the producer and the consumer. It is tempting to think that one offsets the other, but this is not true. When the consumer gains, his gain is less than what the producer is losing and *vice versa*. This is due to many factors, but particularly to the following phenomenon: in the face of risk, producers are careful and take "risk premium" (see Box 2). Instead of basing their profitability calculations on average prices and costs, they overestimate their costs and underestimate their probable profits. This is a deplorable behavior, but which is deeply entrenched in human nature[1]. It is therefore vain to try to thwart it. The consequence is that "all things being equal", the quantity produced is always lower in risk situations than in situations which are "certain". As a result, the overall quantity produced is "on average" smaller, and the "average" price higher. And, with an inelastic[2] demand, the quantity must not be much smaller for the price to become extremely much higher...

> *Price fluctuations would be eschewed if demand and supply bear on quasi constant quantities*

Thus, fluctuations and the risks they generate are deeply wicky. In fact, because they normally result in average price increase, they are especially prejudicial to consumers. However, what producers gain from the "price" is lost in the "quantity", such that in the final analysis, everyone is a loser in the game. In reality, price fluctuations have the same effect as "reverse" technical progress, which reduces the efficiency of production systems. It is therefore important to envisage either eliminating them or, at least, mitigating the effects thereof. However, trade appears to be the proper solution, at least if fluctuations are caused by changes in weather – as most economists think.

---

[1]Although in some cases (as people can win in lottery games!), the imprudent who do not adopt such rules can earn a fortune, most of the time, stakes will be lost.
[2]The notion of "inelastic" or "elastic" demand is conventional in economics (see below, in chapter II.)

Indeed, if such is really the case, and as it is unlikely that a drought or another similar accident may occur simultaneously in Europe, the United States and Australia, one may expect that good harvest in one region would counterbalance bad harvest in another, on condition that trade between the continents is possible. More rigorously, if one assumes that harvest discrepancies are distributed along what mathematicians refer to as "Gauss' law" – the regular "bell-shaped" probability distribution, so familiar to statisticians –, then there is a theorem called "the law of large numbers", which ensures that world production (which is the sum of a large number of small unforeseen occurrences of this kind), would remain virtually constant (see Box 3). Figure 2 illustrates this reasoning. Price behavior in five identical random markets has been traced therein, and

---

BOX 3   THE LAW OF LARGE NUMBERS AND MARKET STABILIZATION

Let us assume that the world is divided into N regions. In each region, production is a random variable of expectation $\bar{p}$ and variance σ^2, the same for all regions. (The reasoning is very similar if the data vary from region to region, but that complicates the presentation unnecessarily.) The probabilities are different from one region to the next, such that the average of the sum is the sum of the averages, and the variance of the sum is the sum of variances. At international level, production is thus $N\bar{q}$, the variance is $N\sigma^2$, and the standard deviation is $\sigma\sqrt{N}$.

What is important is the deviation probability as against the average, which depends on the standard deviation. Thus, by assuming (additional hypothesis) that output distribution is Gaussian, the probability of a 20% deviation from average production in a region would be 0.15. Let us assume that there is a price elasticity of demand of 2.5. (See ahead in Chapter II, the meaning of this notion of elasticity and the justification for attributing such order of magnitude to it.) There would be a 50% price increase, once every six years on average if the regions are isolated from one another. If there are 100 regions (N = 100) that form a "pool" within the context of an international market, then the overall standard deviation would be multiplied by 10 and not by 100, such that the average deviation that would occur once every six years would just be 2%, resulting in 5% price variations.

Such an example is of course barely illustrative. (In practice, it would be normal to consider non Gaussian probability distributions: there is no justification for choosing this special distribution, and there are even grounds for disregarding it, though explaining such reasons here would take too much time and would digress from our main focus. As a consequence, the results would be slightly different.) It suffices to explain the gist of the argument.

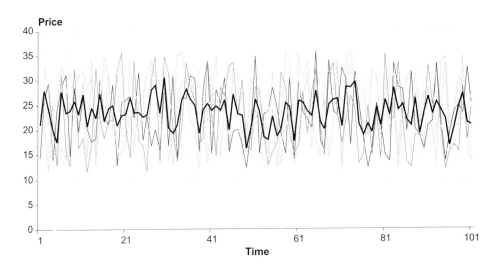

Figure 2 Effect of merging five independent random markets

Five identical markets are defined by a supply curve with a 0.35 slope passing through the origin, and an ordinate demand curve from origin 50 and with a -0.4 slope. The equilibrium quantity is 66. They are subject to random supply shocks, evenly distributed between 36 and 96, and independent from each other. The graphs in the background represent the prices prevailing in these markets. They fluctuate between 10 and 35.

The thick curve at the centre represents the price obtained after merging the markets: the new demand and supply curves are determined in such a manner as to represent the sum of demand and supply in each of the five markets at that price. One thus observes that the price fluctuation range is reduced between 17 and 30: expansion of the market has firmly stabilized it.

represented by the five lean curves. Price fluctuations are caused solely by variations in the quantity produced. For each of these curves, producers planned production is always 66. However, independent random shocks cause actual production to randomly take any value evenly distributed between 36 and 96. Prices are then moving according to the demand curve in face of this actual supply. If the five markets are merged into one and the quantity and demand are added together, one would obtain the prices represented by the bold curve in bold at the centre of the figure. It is visibly less "volatile" than the others. Such is the benefit expected from liberalization.

If international markets were operating this way, demand and supply would be based on quasi constant quantities. There would be no justification for price fluctuations. By contrast, if independent markets, isolated from each other, have each to balance demand and supply under random shocks in every part of the world, then, fluctuations are likely.

According to this reasoning, liberalization would be doubly beneficial: on average, it would enable to exploit comparative advantages; in addition, it would forestall agriculture-specific price fluctuations so prejudicial to the general well-being. Consequently, the case for liberalization seems sufficiently founded. One then begins to wonder why liberalization did not occur earlier! One also wonders why, in the course of history, all the benefits associated with free trade were abandoned, whereas free trade corresponded to the "normal state" of society.

This (apparently innocuous) question implies that there is something fishy going on: except if we consider they were all fools, our ancestors must have had their reasons for changing the state of nature. Authors like Olson (1965) or Gardner (1992) have tried to demonstrate that such change was brought about by agricultural lobbies, which took unfair advantage of their influence to extort from society undue benefits for their own gain, to the detriment of consumers.

There is no doubt that agricultural lobbies were sometimes powerful and in some cases succeeded to secure exaggerated benefits for their constituents. Should we however generalize to say that agricultural policy measures were all useless? Here, we are faced with a problem similar to the French writer Jean Jacques Rousseau's in the mid 18th century. If the "savage man" were as good as is claimed by the author of *Emile* in his theory, then why did he allow himself to be corrupted by society?

In practice, many factors compel man to live in society, and still many more underpin the regulation of agricultural markets. We will now leave the fate of man and society and tackle the issue of agricultural markets.

CHAPTER II

Theoretical Criticism of Agricultural Liberalism

Why these past and present obstacles to free and "natural" trade?

If the pro-free trade arguments are so strong and universal as they seem, how come that throughout the world, and at different periods in history, nations felt the need to erect all those trade barriers here and there? Why are attempts at abolishing them through the organization and reorganization of expensive international conferences so painfull?

To understand this, it is necessary first to consider why, for quite a very long time, great minds have argued against liberalism and found interventionism wanting. The criticism of liberalism is based on two foundations: first, experience which has shown that, especially in agricultural matters, it did not necessarily yield the benefits expected thereof; second, reflection, which is used to interpret experience. Logically, we should begin with experience. But this would waste us time and may mislead us or plunge us into details. That is why we will begin by theoretical criticisms against agricultural liberalism in the light of past experiences. Then in the following chapter, we will return to facts to see how they challenge or support theory.

Criticisms against liberalism are of two kinds. Some of them oppose the very principle of market economy. Paradoxically, they are not very serious because though they highlight very serious objections against the perversions of the system, it is possible to remedy the weaknesses they denounce from within. Other criticisms concern the functioning of some markets under certain circumstances. Such criticisms are more preoccupying because they hit the very ability of the system to operate the way people think it does. Whatever the case, they undoubtedly justify the "agricultural exception", as foreseen by Franco Galiani (see Box 4). We shall begin with the first set of critics.

BOX 4 GALIANI

The criticisms against royal regulations at the end of the reign of the French king Louis XV already aroused controversies about their raison d'être. Galiani (1728–1787) is undoubtedly the most prominent of such critics. His book titled *Dialogue sur le commerce des bleds (A conversation about corn trade)* (Galiani, 1770) has not lost much of its validity for over the past three centuries. He is one of the first to have developed a theory on the idea of "agricultural exception": what is good for industry is not necessarily good for agriculture and *vice versa*. He however did not exert any real political influence, at least at the time: he did not prevent prime minister Turgot from undertaking the liberalization of trade in grain which ended with the 1775 "flour war", which many historians consider as a kind of rehearsal of the French Revolution. Later on, after the French Revolution, during the first half of the 19th century, the institution of the "sliding scale" in France can be attributed to his influence. Such system of variable import levy was not unrelated to the common agricultural policy instituted in 1960. The sliding scale takes up many of his ideas. This also applies to the English Corn Laws policy as practiced during the same period. Galiani was again brought to the limelight by Joseph Schumpeter, who paid him great tribute in his famous "History of Economic Analysis".

A Radical Criticism: Liberalism and Justice

We have seen above that the basic advantage of markets is to ensure that under the influence of competition, the minimum possible prices are fixed in a given technical situation. This point is rarely challenged, at least for "theoretical" markets, that is, those markets described in economics textbooks. (As for "real" markets, whose functioning may be observed in practice, the situation is different and will be discussed below.)

Price, a transmission agent in a likely unequal trade system

As such, price acts as a transmission agent between the producer and the consumer, informing the producer about the desires of consumers and informing consumers about the difficulties involved in production. Thanks to such information, effective compromise is reached between the contradictory interests of the parties involved. But that is not the only thing that price does. Price also distributes incomes. If, owing to price and competition, a factor of production becomes "scarce" because it is needed for the production of a highly demanded product, its holder automatically becomes rich by virtue of his proprietary rights alone. Conversely, the owner of a good that exists in abundance, or is used in the production of another good that is not highly demanded at a given point in time, automatically becomes lowered to the category of the poor.

Thus, the proprietor of a diamond mine is owner of a fortune which has cost him nothing more than the pains of birth, notwithstanding the fact that diamond is not in the least indispensable to life. By contrast, the poor water boy in an Asian town supplies a really crucial good for the survival of his fellowmen. Yet, he leads a miserable life because his only asset is his unskilled labor, which is abundant in those countries. Is it then "fair" that the diamond trader should wallow in wealth while the water boy leads a miserable life? The answer to this question is a problem of moral and not of economic analysis. Yet the economist cannot avoid asking himself this question because, inevitably, a negative answer would imply "distortions", which are not necessarily unjustified. This is all the more true given that within such context, "misallocation" of wealth may be self-sustained: diamond is an object of luxury, which only has value because some persons are rich enough to demand it though not having vital need for it. Such demand at the same time creates the wealth of the diamond trader...

Is it therefore possible to allow free trade and market forces to blindly regulate such important issues, which so closely affect every citizen? Many analysts and prominent ones have provided negative answers. Contrary to what could be seen, however, such objection to free trade that may appear fundamental is not also as dramatic as one may think. This is due to the fact that solutions may, at least theoretically, be designed to maintain market efficiency in allocating resources (this is referred to as "allocative" efficiency), while at the same time correcting their negative effects on income distribution through an appropriate taxation system. To achieve this, it suffices to institute a suitable taxation system which superimposes itself on free trade without perturbing it. In this way, "exaggerated" profits would be automatically confiscated, and the balance would be the fair compensation of effort and talent to the benefit of all[1].

No global taxation system to correct inequality

Naturally, such a taxation system is not neutral vis-à-vis production. Meeting the needs of a limited number of rich people within a miserable population and meeting the needs of an egalitarian population implies different approaches. The volume and nature of production cannot be the same in both cases. But what[2] ever the taxation system, if markets are free and well behaving, one can be sure that the resources would be used "efficiently". This explains why the market is not necessarily incompatible with "social justice", on

---

[1] Moreover, concerning the agro-industry, such systems were set up a long time ago, particularly for military purposes. For instance, Daviron (2003) and Daviron and Voituriez (2003) show how during the last world wars, the belligerents developed food card systems and other food stamps. The main function of such systems was to correct the negative impacts of the free market on the feeding of the "poor" during periods of great uncertainty. They were necessary to prevent hunger in the rear that could demoralize the troops at the front.

[2] For instance, with a very unequal income distribution, 100 Rolls Royce and 100 000 bicycle will be produced. With a more even distribution, the market will require producing 50 000 medium sized cars. But bicycle, Roll Royce and medium sized cars can be produced efficiently or not, whatever the distribution of incomes.

condition that the profits are distributed through a suitable taxation system.

Also, a distribution system that can correct the undesirable effects of liberalism does not develop spontaneously, and instituting one is no trifling matter. With regard to international trade in particular, one must admit that there is no global taxation system, and that consequently, there are no obstacles to the failings of "laisser-faire". This paves the way for what was incorrectly referred to in the 1970s as "unequal trade": this is not a system in which the poor are "exploited" (they may avoid being exploited by refusing trade), but one in which they, in any case, get just the crumbs of the profits confiscated almost entirely by the "rich". This may be considered as "natural", which of course is so, just like disease or death. However, this does not mean that such a situation is acceptable.

It is important to assess the extent of the problem and to reflect on the remedies thereof. Such may be the use of econometric calculation "models" that will be discussed later. But before we delve into a description of these models, it is necessary to further discussion on the limits of the preceding analysis.

Another Radical Criticism: Markets and Needs

If we assume that the preceding problem is solved, it would still be important to know whether all individual and collective needs are likely to be met by the market. In effect, theory tells us that if any solvent need arise, there will be an entrepreneur to notice it, and do everything to respond to it. Such response must be to the satisfaction of both the entrepreneur who can make profit, and the happy customer who get pleasure to consume. Yet, this is not true in every case. Indeed, for such an arrangement to occur, the customer who is paying should have to be technically the sole beneficiary of the entrepreneur's effort. If not, this would give rise to the problem of the free rider who may take advantage of the good without paying.

Agricultural prices influence the environment

In the case of agricultural products which is of concern here, situations of this kind are frequent. For instance, no one can deny the pleasure derived from watching a beautiful landscape. But a beautiful rural landscape is not natural. It is something designed through farmers' decisions about rural building, trees, roads (also used to transport agricultural produce), etc. Obviously, such decisions depend on the interaction between natural conditions and farmers economic environment. Consequently, any price change is likely to change the landscape in a way that users may find "disastrous". At the same time, the market does not provide the latter with any means of alerting the farmers about the degradation of their situation. That is why there are laws on environmental protection. However, the way these laws are implemented can modify the "comparative advantages" and lead to "distortions". It is very difficult to manage this kind of contradiction within a fully liberal context. This is why even a body like the Organization for Economic Cooperation and Development (OECD), which cannot be suspected of socialism, recognizes the multifunctional nature of agriculture.

An Apparently Incidental Criticism: do Markets Function?

The preceding remarks regarding the benefits of liberalism presuppose, in any case, that markets function well, that is, competition inevitably causes price to be situated close to the minimal production cost feasible. Unfortunately, this is highly questionable, particularly with regard to agricultural products, because of price fluctuations.

In the preceding chapter, we discussed the negative effects of such fluctuations, which liberalization seems likely to mitigate. However, for such mitigation to occur, the causes of fluctuations must be outside the pricing system. They must be due to the many random shocks, for instance relating to climate, each of low amplitude.

But there is another much more disturbing theory relating to the fluctuations of commodity prices, which links the cause of fluctuations to the functioning of the market.

> *Agricultural marks are unstable*

In effect, a market is a dynamic system: at all times, "shocks" are produced and destabilize equilibrium. For all systems of this kind, the effect of a shock is not serious if the system is "stable", that is tends to return to equilibrium when destabilized. Such is the case of a ball placed at the bottom of a cup: the cup may be shaken to destabilize the ball. Unless the shock is excessive, the ball would soon return to rest at the bottom of the cup until another shock comes round. This is not the same with a ball located at the tip of a pen: formally, it is in equilibrium and a juggler can maintain it in position. However, in the absence of the juggler, the slightest breath would cause it to fall off, thus destabilizing it definitively from its position of equilibrium, which it was able to maintain just for a few seconds.

Market stability is governed by what economists refer to as "demand elasticity". Economists are unanimous (once does not constitute a habit!) in stating that the demand for food products is inelastic, and therefore "rigid". Decreasing the prices of food products would not cause people to eat more. Similarly, increasing such prices would not cause them to eat any less. Obviously, this does not apply to any particular product. If the price of chicken drops as against the price of beef, demand for chicken would surely increase while the demand for beef would drop. Halving the price of caviar would lead to a tenfold increase in the demand thereof. But all of this is done by substitution of one product for another. The overall quantity of proteins and calories consumed would not be much affected by such price fluctuations. Generally, therefore, demand for food is inelastic.

The consequences of this observation are far-reaching. First of all, it justifies the practice of treating food problems as one whole – without considering beef-related problems apart and those of chicken separately. A low demand elasticity, especially, causes instability in agricultural markets.

In fact, when the price of a product increases, demand decreases: this is the law of supply and demand. But the situation is very different depending on the extent of the decrease in sales, in response to a given price increase. If a 1% increase in price causes demand to decrease by more than 1%, such demand is said to be elastic, and the producer's income decreases with the increase in supply. This kind of situation is obviously favorable for the re-establishment of market equilibrium. Conversely, with inelastic demand and a price increase of 1% too, demand will, for instance, drop only by 0.5%, and the seller's income will increase with the rise in prices, which would not, *a priori*, allow to re-establish equilibrium. Figure 3 below illustrates this phenomenon.

This observation was made more than three hundred years ago by Gregory King, who was studying the functioning of the London Grain Stock Market, one of the first commodity markets in history[1]... King came to the

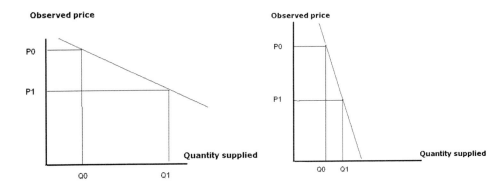

Figure 3 Elastic demand (left): a slight fluctuation in supply leads to a moderate change in price. Inelastic demand (right): a slight change of supply leads to a significant change in price.

---

[1] He observed that when supply at the London Stock Exchange decreased by 10, 20, 30, 40 and 50% below the "normal" level, prices soared by 30, 80, 160, 280 and 450% (Schumpeter, 1954). This corresponds to price elasticity of demand of about 2.5.

conclusion that farmers earned more money during years of "poor" harvest than during years of "good" harvest. This was already something that fired the imagination of those who bank on personal interest to meet general needs: as a rule, it is in the interest of farmers to organize shortages!

Fortunately, competition prevents them from doing so because while it is in their "collective" interest to have shortages, they individually need to sell as much as possible... Indeed, "sell as much as possible" is not even the appropriate expression: to earn money, the cost of production must not exceed revenue. Therefore, once cost becomes very high, one has to stop... But how can it be possible to compare costs and revenues if one is not sure of prices?

Here lies the entire significance of market instability: the producers having to speculate on the prices that would be prevailing at the time of supply cannot exercise their talents under very good conditions.

Expectation Errors and Endogenous Price Fluctuations

At sewing time, no farmer can tell at what price the sewed product will be sold. Yet, sewing (and well as all kinds of investments) is made in the hope (expectation) that the price will be remunerative. To understand why and what a farmer plants, one has to understand how are expectations built up and beliefs regarding futures prices made.

BOX 5   THE NOTION OF DEMAND ELASTICITY

The notion of demand elasticity is one that is on the mind of the humblest retailer. Every chip seller asks himself: "If I reduce my price, would I counterbalance the income loss on every object sold through an increase in the quantity sold?" For this to happen, sales have to increase by more than 1% when price drops by 1%. Price elasticity of sales is the ratio of the relative increase in sales to the relative increase in price. Such elasticity demonstrates the capacity of the market to "respond" to price: a low elasticity indicates that demand is inelastic and not sensitive to price

> fluctuations. In such a circumstance, the chips seller referred to above may reduce price, but will not increase sales accordingly. Conversely, increasing price will not reduce sales in proportion, thus leaving the possibility for the seller to make enormous profits (Fortunately enough, in a competitive market, competition will rule this possibility out). By contrast, a high demand elasticity indicates that demand is responsive to price changes. If I increase price, I would sell much less: this is quite dissuasive for a merchant. Under such conditions, it is not surprising for a market to work all the better as supply and demand are elastic: if demand is inelastic, price cannot be used to regulate demand and supply, because whatever the case, the customer would buy. Similarly, if supply is inelastic, there will be need for very elastic demand to determine an equilibrium price that would neither be ridiculously high nor exceedingly low.

Many "standard" economic models – particularly those that in the recent past have been used to calculate the benefits to be expected from liberalization – posit, in this regard, that farmers anticipate what would be the equilibrium price on the market. This assumption is referred to as "rational expectations" because it would indeed be rational to proceed this way[1] if it were possible. Unfortunately, there are no grounds for affirming that things effectively happen this way.

Price fluctuations are chaotic

The consequences of the rational expectations hypothesis are, however, not necessarily as regrettable as may be feared. If it happens that prices spontaneously return to equilibrium when displaced therefrom, then this hypothesis is clearly justifiable, provided that one is dealing with long-term phenomena: even if operators make a few errors, the system would spontaneously shift back to equilibrium. That's why many economists so easily accept the idea, which in addition facilitates computer calculations. But what would happen if equilibrium is unstable, as we have just seen in the case of agricultural produce?

---

[1] In all rigour, the "rational expectations" hypothesis does not say that agents know the equilibrium price but that they "rationally" process the information available when they are making their decision, which is probably acceptable. In reality, it is by (abusively) introducing a second hypothesis – rational processing of available information would enable agents to guess the equilibrium price – that one is able to justify the common practice of "general equilibrium models".

What happens in this case is potentially very regrettable, as demonstrated by one of Franklin D. Roosevelt's advisers, Mordecaï Ezekiel, at the end of the 30s. His demonstration is known as the "*cobweb theorem*, though here there is neither theorem nor spiders. Let us assume that (which is plausible) farmers consider the current price as the equilibrium price on the market. Let us again assume that for some fortuitous reason, such price corresponds to a quantity supplied q_0, which is much smaller than the equilibrium, as indicated in Figure 4. In this case, the price rises up to point P_1, which corresponds to q_0 on the demand curve – the curve that matches the price on the axis of ordinates with each quantity supplied on the axis of abscissas. At this price, producers are optimistic: they increase costs to raise supply up to q_1, the point where price is equal to cost, as represented on the "supply curve" – the curve that expresses the relationship between the cost and the quantity produced. But it is difficult to find buyers for such a quantity: the price then drops to P_2. At this price, production decreases to q_2 and so on and so forth.

Clearly, price rises and drops indefinitely... But that is not the end of the story yet! The magnitude of fluctuations is not constant: with such system as one sees in Figure 4, under certain conditions, fluctuations are mitigated with the passage of time. In technical terms, it is said that they converge.

After a number of periods, fluctuations become imperceptible and the price stabilizes at equilibrium. In this case, is everything not working for the better? Of course not! Things work well in some "cases" but not in all. Let's try to tilt the demand curve on the axis, to make it "inelastic". One can easily see that instead of converging, the system is "exploding". Fluctuations keep worsening with the passage of time to the extent that one soon reaches negative quantities and prices... Yet, we have seen that agricultural demand is inelastic. Should one then expect sometimes excessive agricultural prices and sometimes negative prices?

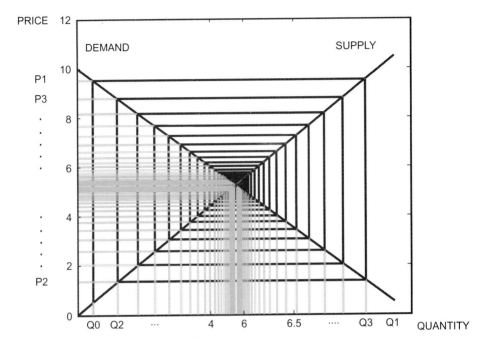

Figure 4 The cobweb model. Starting from quantity q_σ we obtain price P_1, which prompts the supply of q_1, the quantity whose equilibrium price is P_2 at which q_2 is supplied, etc. Here, because demand is elastic, the system "converges" towards the optimal equilibrium.

Naturally, this is absurd! Something has to bring the price close to equilibrium in such a situation. In fact, there are several mechanisms likely to achieve such a result. Among them is one that was developed over a century ago[1] by the Swedish Knud Wicksell to explain fluctuations in general activity: when prices and quantities are obviously very far from the equilibrium, operators panic. In order to continue to produce, they demand "risk premiums[2]", which are much higher profit

---

[1] It is in 1898 that Knut Wicksell published in German his famous book titled *Geldzins und Güterprise*. It was translated into English much later, in 1936, under the title *Interest and prices* (Wicksell, 1898).
[2] See Boxes on risk premium (Box 12, Chapter VII) and on the Markowitz model (Box 11, Chapter V)

rates than in normal time. Thus, instead of increasing their production inconsiderately as "market forces" are urging them, they reduce it in the hope of reversing the situation. Price and quantity are then brought close to equilibrium. Given that this position is unstable, they cannot sustain it: the cycle then resumes indefinitely.

BOX 6 ALGEBRAIC EXPRESSIONS OF THE COBWEB THEOREM

The Traditional Cobweb

If one assumes that the demand and supply curves are linear, then the system of equations below would be the algebraic expression of the reasoning presented in the text on the cobweb theorem:

$$\hat{p} = p_{t-1} \qquad (1, \text{expectations}),$$

$$p_t = \alpha\, q_t + \beta, \qquad (2, \text{demand}),$$

$$\hat{p}_t = a\, q_t + b, \qquad (3, \text{supply})$$

where p_t and $\hat{p}_t$ are respectively the real and the expected prices at date t; α is the slope (negative, so much greater in absolute terms as demand is more inelastic) of the demand curve; a is the slope of the supply curve, that is, of the rising marginal cost; b and β are constants.

One can easily see that when t increases, the system converges towards an equilibrium if $|\alpha/a| < 1$, or diverges (the prices and quantities become infinite) if $|\alpha/a| > 1$, and remains periodic for $|\alpha/a| = 1$.

In the case of agriculture, there are chances that owing to inelastic demand, that is a high α value in absolute terms, a situation of "divergence" may occur. As prices have never become infinite, there is somewhere a need for a call back mechanism that obliges the system to return close to equilibrium whenever it has moved too far from it. It is possible to develop many mechanisms to this end. However, the simplest – and which undoubtedly better corresponds to what can be noted from discussions with producers – is no doubt the existence of risk, and the methods used by farmers to avoid it.

The "Risk Cobweb"

One traditional economic method of taking risk into account consists in assuming that decision-makers determine a weighted average of their "profit expectation" (the average profit) and of the risk associated with such profit, measured by its variance (this idea is discussed above in Box 2 on risk aversion). In the sketchy model presented here, the average value of price expectations is assumed to be invariant:

$$\hat{p}_t = p^0, \qquad (1\text{ bis})$$

where p^0 is a constant that may be equal to the market equilibrium price, $p^e = (a\alpha - \alpha b)/(a - \alpha)$, or differs therefrom. But the producer does not know p^e precisely. Thus, a risk premium is taken. As we have seen above (see Box 2), the first order condition for utility maximization is then:

$$\widetilde{p}_t = \hat{p}_t - Aq_t\hat{\sigma}_t^2 \qquad \text{...(4)}$$

where $A\hat{\sigma}_t^2$ represents the expected variance of the price for the period t and A is a "risk aversion coefficient". To close the model, there is need for an equation to define $A\hat{\sigma}_t^2$. Something in the spirit of naïve expectations could be:

$$\hat{\sigma}_t^2 = (p_t - p_{t-1})^2 \qquad \text{...(5)}$$

With these specifications, the dynamics of this model become really interesting. Figure 5 gives an idea of the type of result obtained with relatively plausible values for the parameters. Naturally, one gets different results for other values of the parameters and the model still converges if $|\alpha/a| < 1$ (Figure 6).

One may develop many variants of this kind of systems, for instance by taking capital accumulation into account (see, for instance, Abraham-Frois [1995]).

Figure 5 shows the fluctuations that would occur with a slightly modified *cobweb* to take into account the preceding remarks (Box 6).

This figure is all the more striking when compared with Figure 6, constructed with exactly the same equations and the same model, except that the "demand curve slope" – in other words, the inelasticity of demand

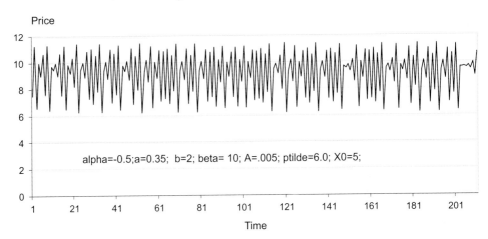

Figure 5 A "chaotic" cobweb. This figure is constructed with the equations indicated in Box 6 as follows: $\alpha = -0.5$; $\beta = 10$; $A = 0.005$; $a = 0.35$; $pe = 6$.

- was changed. Whereas the demand curve that was used to construct Figure 5 was "inelastic" because strong variation in price did not very much change the quantity demanded, in Figure 6, it is "elastic" because any price variation significantly changes demand. As we have seen above, in this case, the *cobweb* "converges" and the market returns to its natural equilibrium even if it has shifted therefrom.

It is important to understand the profound difference between the latter fluctuations and those discussed in the preceding chapter. The random fluctuations we discussed were "exogenous": they were due entirely to climate or to other fortuitous circumstances, independently from the pricing system. In addition, their sources were independent of any of the measures taken to mitigate their effects.

This explains why it was possible, without changing the sources or magnitude of the independent shocks leading to fluctuations, to eliminate the effects thereof by simply adding together supplies from different regions. Here, the situation is different: the fluctuations we are

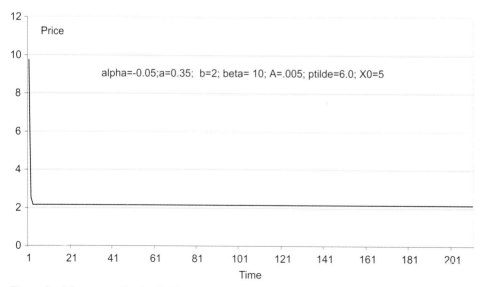

Figure 6 A "convergent" cobweb. Constructed with the same formulae and parameters as in Figure 5, this figure is only different in that the demand curve slope moves from - 0.5 to – 0.05.

> Markets do not regulate fluctuations.

studying are "endogenous", created by false expectations and an unstable equilibrium, which are intrinsic characteristics of the market. Any attempt at mitigating consequences would change the model parameters and have unpredictable effects.

For instance, on Figure 7, results of the preceding *cobweb* model applied to two "markets" are shown. Markets a and b are independent from each other in the upper part of the figure, and interrelated in the lower part[1]. A study of the curves of the upper part, with one market going downward while the other is soaring, strongly suggests encouraging mutual communication between the two markets as a convenient regulation policy. In this way, excess in one market would counterbalance shortages in the other. Alas, this is an illusion, as shown by the curves on the lower part of figure 7: with this kind of endogenous fluctuations, if the quantities sold from one market to another increase, the two markets simply synchronize (with curves which are so well superposed that it is no longer possible to distinguish them), but fluctuations persist. This is quite different from what we got in the preceding chapter (Figure 2).

It is difficult not to relate these theoretical and "artificial" figures to those of completely random fluctuations which are very much similar. It is also difficult not to relate them to real fluctuations of agricultural prices, such as those presented in the preceding chapter (Figure 1). In fact, fluctuations of tomato prices may be interpreted as those of a sort of *cobweb*, though less regular than that of a chaotic model but nevertheless quite similar.

---

[1] In fact, they are neither completely linked nor completely independent. However, each supplier is compelled to sell quantity C of his production on the other's market, irrespective of the price; $C = 0.1$ in the upper part of the figure, $C = 0.3$ in the lower part of the figure. For the rest, the parameters of the two curves in each graph are identical.

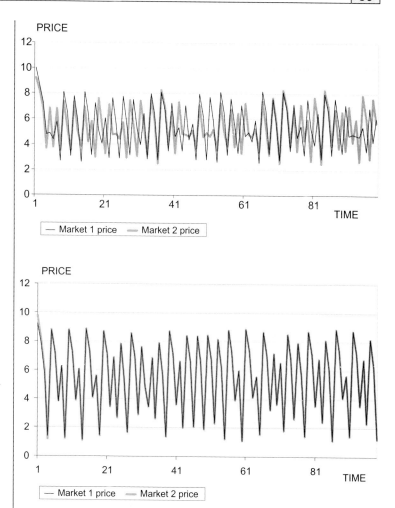

Figure 7 Merging two chaotic markets: independent markets (above) and related markets (below)

Here, we use the same approach as in Figure 2, but with only two markets. The markets are chaotic cobwebs as before, with identical parameters and slightly different starting points. They communicate: for the upper graph, 10% of the production of one market is sold on the other market; for the lower graph, it is 30%. We see that limited communication (upper part of the figure) gives rise to independent fluctuations of the two markets because of different starting points. Conversely, with close communication, (lower part) fluctuations in the two markets, far from offsetting each other, synchronize. The two markets become merged into one which continues to fluctuate.

In the preceding chapter, we discussed the negative consequences of fluctuations. It is therefore inevitable to conclude that markets do not function the way they should. This casts doubt on the validity of preceding analyses concerning the benefits of liberalization. Then, what to conclude in practical terms? How may the problem be tackled from the point of view of quantity as was the case in the preceding chapter? To address these issues, it is necessary to revisit the design of the model that was used to get results presented in Chapter I. Modification of the model may then be envisaged in such a manner as to incorporate a few of the considerations that have just been expressed. We would then see whether such remarks are really important. Perhaps, after all, though not irrelevant, they do not challenge previous conclusions too much?

CHAPTER III

The Test of Facts

The free market, a natural institution

Theories would not be of much importance if they were not rooted in experience. From a strictly theoretical point of view, just anything and its contrary can be imagined. Since the beginning of time, economists too have done this.

It is interesting to note that pro-protectionism theories, which have just been summarized above, rarely ever developed spontaneously. The theories that "naturally" come to mind are those of Chapter I, which favor liberalism. Other systems have been developed in reaction to the principles of liberalism, because the latter did not take into account what actually happened. This is why we need to start with some little historical background.

Brief History of Liberalism

Contrary to a currently widespread opinion, globalization and world trade have a longstanding history. Archaeologists have found traces thereof dating back to the Paleolithic period, with the discovery of firestone or other materials thousands of kilometers from their original deposit. These objects must have been transported and, very likely, traded.

> *Whoever controls trade controls the country*

In reality, trade is quite natural to the human mind. The "free" market is a "natural" institution that develops spontaneously without need for any political will. The phenomenon is so obvious that, paradoxically, writers and philosophers only belatedly raised concerns about its growing importance and its relevance. It is the fruit of such reflection that has been reported in the preceding chapters.

However, this is not the least paradox of economic history. The most profound reflections on the benefits of free trade occurred at a time when generalized systems of customs and trade barriers were being set up. Indeed, before the 18th century, states were generally too weak to establish real barriers along their frontiers. In addition, as Paul Bairoch has aptly demonstrated, transport costs at the time were so high as to prevent trade of any ponderous commodity: if silk or Chinese porcelain were transported, it meant that such products had a high value per ton or per cubic meter. Accordingly, trade was reserved for luxury products, for "superfluous" goods, according to Jean-Jacques Rousseau[1]. Transportation of grain under the same conditions was meaningless. Trade therefore did not involve "basic commodities"[2]. In reality, it is the advent of trade in "basic commodities", owing to improvements in the means of transportation, that gave rise to the issues that are still raised today.

In fact, the development of transportation facilities and routes brought into competition economic systems that were hitherto not competing. The practical disadvantages arising from such a situation then emerged and triggered

---
[1] This author's contribution to economics is generally unrecognized. Yet, one may wonder whether his analysis of trade does not by far outmatch that of most "writers" of that period.
[2] With, undoubtedly, a notable exception, namely supplies to Rome during the imperial period. Again, this was not yet real free trade, for emperors had control over the transactions. One could even conjecture that the relaxation of Roman bureaucracy on this issue after the 4th century contributed to the dissolution of the empire.

off pragmatic reactions often motivated by emergency (and not at all by any theory). This explains why, for instance, the authorities of royal France were very quick to express concern about the disadvantages that could arise from inconsiderate exportation of buffer stocks, and the political disasters that famines could bring about. In reality, in the 16th and 17th centuries, the organization of domestic trade in food products was one of the signs that the royal authority was capable of controlling the country.

In the 18th century, a paradox emerged and is still lingering today: the benefits of such organization were often no longer visible because, owing to the efficiency of the system, the disasters of the previous centuries were already forgotten. Famines (some cases still persisted) were attributed to wars (this happened often) and not to markets, which were already beginning to build-up and whose success was celebrated by contemporaries with astonishment. At the same time, the disadvantages of royal bureaucracies, often led by routine-minded, formalist and mediocre personalities were very perceptible. One could not transport a wagon of grain from Limoges to Paris without obtaining authorization from the District Administrator. The latter delayed his decisions and sometimes demanded spices in return. It is clear that this kind of situation neither favored trade nor the fight against poverty. This probably explains the growing importance of "liberal" arguments of the time and the emergence of physiocrats.

The French were not the only ones in this situation: England faced the same problem with its *Corn Laws*. Indeed, this generated the debate in which Ricardo participated. Given that the latter's argument was very seductive, it prevailed over the others and England liberalized. On the continent however, thinkers were divided. Throughout the 19th century, there were attempts at liberalization immediately followed by upsurges of protectionism.

In France, the beginning of the 19th century, from 1790 to 1815, was characterized by militant protectionism, which at least caused the fall of Napoleon[1]. Within France, trade was in principle free, supervised in practice by imperial authorities.

The period of recovery after Napoleon was characterized by moderate protectionism and the "sliding scale" policy – in fact, this foreshadowed the import tax/export refund system of the European Economic Community already proposed by Galiani and, besides, also enforced in England.

Protectionism or free trade depending on opportunities

In the rest of Europe during the same period, the debate on protectionism was very much animated – like today –, in particular because German unity was centered around the *Zollverein*, the customs union of German principalities, which is not unlike the approach adopted today by the European union. The theoretician behind the approach was Friedrich List, a German who emigrated to the United States (after a brief stay in Paris) and who came back to Hamburg as the United States Consul. Unlike Galiani, he advocated industrial protectionism, but remained committed to agricultural free trade. His main concern was German development. To achieve such development, there was need for a competitive industrial sector, which could only be based on low prices for grain, the basic food stuff. Since cheap grain used to come from the United States, a new country where land was cost-free, grain was to be imported from there. Conversely, development required that the budding industries be protected against undue competition with the English entrepreneurs who did not hesitate to resort to dumping to eliminate rivals: it was therefore necessary to protect the German industry against unfair competition. A powerful argument in this line was that the United States

---

[1] Indeed, it is in an attempt to enforce the "continental blockade" that he attacked Russia (which was a liberal State and intended to export its grain to England that did not produce enough).

had already practiced such protectionism with outstanding success...

Although Friedrich List's work is serious because the author masters his subject, it is all the same theoretically mediocre and cannot compete with Ricardo's flamboyance. This is perhaps the reason why he is attributed little esteem by economic theoreticians. Nevertheless, and notwithstanding his suicide connected with the fact that he was misunderstood, Friedrich List had considerable practical influence: he inspired Bismarck who imposed his protectionist policy on all Eastern Europe after 1870, particularly in France.

In the meantime, England had resolutely become a free trade economy increasingly counting on grain from the United States to guarantee her food security. This was perhaps the cause of the great famine that occurred in Ireland in 1848. Yet, at about that time, the English sought to export their ideas on free trade to France under Napoleon III. The latter, perhaps to obtain pardon for its internal dictatorship denounced by Victor Hugo, undertook to align with the Anglo-Saxons[1] in his external relations. However, after 1870, such vague desire did not survive the advent of the Republic under German domination.

In France, England, and to a limited extent Germany and the United States, the end of the 19th century was marked by the "colonial pact": free trade limited to the colonies, based on the obvious complementarity between the colonies that exported agricultural commodities and imported industrial goods, and the metropolis which did the reverse. This was a direct application of Ricardo's

---

[1] Such policy struck contemporaries, sustained conversations and influenced behaviors: Bruno Latour (a French sociologist and historian of political ideas), recently exhibited a letter written by Louis Pasteur (the famous biologist and founder of microbiology) to the then Minister of National Education. The scholar requested a subsidy to enable him pursue his work on alcohol fermentation. He justified his request by insisting on the importance of technical progress in oenology at a time when trade in agriculture was being liberalized...

doctrine, though in an area of political influence. Also, colonial trade was strongly regulated, which made it possible to eschew some of the disadvantages connected with the poor functioning of markets. This notwithstanding, it is difficult to affirm that the "colonial pact" was a complete success.

Later on, there was a vague desire to clear customs barriers during the 1920s, in the euphoria following the end of the First World War. Such attempt came to an abrupt stop following the great economic crisis of 1929, which saw the galloping return of protectionism everywhere, even in its most displeasing forms. In the post-World War II period, actions were much more prudent and gradual, with successive negotiation rounds. In the beginning, there was consensus to leave agriculture out of the negotiations. However, since the Uruguay Round, the agricultural sector is at the centre of the problem, given that almost all the other sectors have been liberalized.

One may therefore wonder whether the current trend to liberalize everything is not more of a transient fashion effect than the result of profound knowledge in social arithmetic. In short, it should just be one of those frenzies that from time to time seize political leaders before reality brings them round to more reason. Such interpretation is largely suggested by an analysis of facts concerning two key aspects that distinguish the different pro-agricultural liberalization theories: for some, it would be possible to leave out price fluctuations in calculating the benefits of liberalization; for the others, expanding the market would precisely permit price stabilization.

Price Fluctuations and Growth

We have seen that in pro-liberal discourse, price fluctuations are largely disregarded, except to ensure that liberalization would reduce them. By contrast, anti-liberalists lay emphasis on the fact that fluctuations are

brought about by the market itself, and that they play an essential role in determining supply. To decide between the two schools, two points call for consideration:
- are these fluctuations important in making decisions on production? - The answer is unambiguously "yes", and we will see why;
- are these fluctuations engendered by random phenomena external to the system or are they a genuine consequence of the market itself, in such a way that no market solution can eliminate them? - Economists do not agree on this point.

Price Fluctuations and Production

According to agricultural economists, the role of risk in producers' decisions is curiously ambiguous. On the one hand, everybody says it is important. On the other hand, it is generally very difficult to determine this role and to measure its effect on global economy.

Producers are risk averse

The importance of risk in producers' decisions emerges first of all from producer surveys: when questioned, it is rare for them not to allude to the uncertainty of the future and the difficulty of making decisions under such conditions. However, sociological investigation is difficult[1]. One is never very sure that answers are sincere, particularly in an area like this one which is difficult to grasp because of the concepts at play, which are not so obvious[2]. It is therefore difficult to draw many conclusions from these kinds of sources.

Farmer behavioral patterns constitute other reasons for believing in the importance of price risk. Fifty years ago,

---
[1] Pierre Bourdieu, a celebrated sociologist who died recently, and who began his career in rural sociology, even spoke of "combat sport". More seriously, the major difficulty here is to correctly interpret the answers provided by those concerned. A good example in this regard is provided by Morlon (1987).
[2] Consequently, one obviously cannot ask a farmer "have you an idea of the price variation of Irish potatoes?", when he is well aware of the price of the previous week and knows that such price is speculative.

an American researcher, Rudolf J. Freund (1956), sought to know what could be the optimal cropping system for South Carolina farmers, given the prevailing prices. He used a revolutionary calculation technique, "linear programming", which made it possible to find a solution to simultaneous equations, and which was destined to have a bright future. The solution of the problem thus processed by computer was effectively very different from what farmers of North Carolina practiced: at the time, instead of maize and potatoes that constituted the bulk of local production, the model recommended that nearly the entire region should concentrate on "cattle" and "autumn cabbage".

Yet, Freund thought that it was not by chance or foolishness that farmers chose a cropping system very different from the one that would have enabled them to maximize their gains. There must be reasons. He thought of risk, which, indeed, was greater for cabbage and cattle than for maize. In effect, at the time, cattle and cabbage were not much supported by American policy, while corn was granted a guaranteed price. Freund repeated his model, using Markowitz's specifications (see Box 11, Chapter V) to take the effect of risk into account. The result was an optimal cropping system very similar to real practice.

Price variability confuses market signals

Since then, Freund's experiment has been reproduced thousand times in all countries of the world with the same success. It demonstrates two things: firstly, that it is impossible to conveniently reproduce farmer behavior if one is not aware of the risks they run; and especially that risk, which differs from crop to crop, influences agricultural supplies at least as much as the average price level. This latter consequence is extremely important. We have seen above that the role of price in a market economy is to convey information from the producer to the consumer on the difficulties involved in the production process; and from the consumer to the producer on consumer desires. But the average price level is not, or not only, the sole price property that play this role. Price variability (or volatility, as it is currently described) is also very important. In addition, average levels and volatility do not necessarily move in the same direction. Let us assume that the price of corn doubles suddenly after a long period of stability. Farmers may draw two contradictory conclusions from such situation: they may consider the average price of corn is actually increasing, and this would cause them to increase their production in response to the market signal; they may also conclude that average price remains unchanged, but that price volatility has increased suddenly, causing them to reduce production of a product that has become "dangerous". Obviously, while the first behavior can "restore market equilibrium" (increased production would bring back the price to its long term equilibrium), the second, by contrast, can aggravate the shortage.

This example shows that the major disadvantage of risk is that it "confuses market signals". The latter may be wrongly interpreted, and bring about a situation worse than that which had to be corrected. This is a very serious criticism against market economy. However, if the markets function so poorly, it should be perceptible at the level of macroeconomic data. With the current econometric means of investigation, it should be possible to highlight such phenomena through statistical tests.

Strange enough, this is not all that simple. Most studies that have been carried out in this respect attribute to price volatility only very limited or negligible influence on the overall volume of agricultural production in a country or region. One of the reasons underpinning this phenomenon is that statistical tests are probably based on assumptions that differ from the reality of farm management. The statement to be tested is: "price volatility reduces (stability increases) the quantity supplied of any agricultural commodity". But the assumption generally tested in statistical studies is: "the average of last year's prices" or the "volatility measured last year" influences "production this year". Indeed, this latter assumption is much too precise to be verified. Moreover, volatility is a long-term notion, with slow adaptations. Under such conditions, it is not surprising that assumptions that entail a very rapid reaction by decision-makers are not verified.

Stable agricultural prices increase output

To take this phenomenon into account, it is necessary to carry out qualitative tests with prices classified as "stable" or "unstable". Here, we find that there are significant differences between average production growth rates in different categories of series (Gérard and Boussard, 1994).

Consequently, price stabilization is of fundamental importance, especially to boost production in developing countries. Figure 8 illustrates this phenomenon in the case of the United States and France. In both countries, the output trend curve shows a very net upturn at the

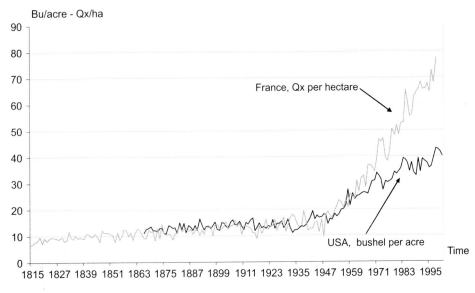

Figure 8 Evolution of wheat yields in the United States (in bushels per acre) and France (in quintal per hectare), from 1815 to 1995

time of instituting producer-price support systems[1]: from 1935-1936 in the United States and from 1945 in France. This is explained by the fact that although the *Office du blé* (Wheat Board) was set up in 1937, farmers only started "believing" in it after the war. The changes in curves slopes are impressive and do not require any specific statistical test to be ascertained[2]. Naturally, it is still tempting here to think that these changes are due to technological progress, which is true. But there was technological progress in the past, which did not give rise to such spectacular results. The pace of technological progress has changed at the same time as stabilization policies were being implemented. This is perhaps a coincidence but if we recall that technological progress does not fall from heaven and that it is "induced" by the economic environment, as was aptly demonstrated by authors like Hayami and Ruttan (1971), we cannot help seeing a cause-effect relationship therein.

Are Price Fluctuations Endogenous?

We have seen above the importance of the stake: if price fluctuations are "exogenous", provoked for instance by climatic variations, then liberalization should have a beneficial and damper effect. If they are endogenous, or created by a cobweb system, then the only chance of reducing them is to isolate agriculture from the market, as was decided by President F.D. Roosevelt in 1935. Unfortunately, it is very difficult to decide on any of the options.

---

[1] These policies were implemented during the "great depression" around 1935 in the United States and in 1937 in France, with the setting up of the *Office du blé*. However, in the latter country, they were effective only after the end of the Second World War.

[2] Obviously, all the statistical tests in question detect a break!

> *Can the chaos caused by the system be tested?*

Many mathematicians have challenged the creation of statistical tests for the existence of "chaos"[1]. Since these series all look like "random" series, it is extremely difficult to distinguish between them. There is a general tendency to highlight "sensitivity to initial conditions". However, since these initial conditions are themselves random conditions, the random element is somehow found in statistical tests.

This notwithstanding, and even if they are not faultless, tests have already been developed. Their application to series of prices of agricultural commodities generally leads to qualified conclusions. One cannot exclude the possibility of these series being subject to chaotic dynamics though one may not categorically affirm that such is the case[2].

Are Price Fluctuations Caused by Climate?

> *Are meteorological phenomena determining?*

In other words, can it be said that agricultural prices fluctuations are only caused by unpredictable events alien to the system, particularly the climate? Here again and strange enough, empirical studies are quite rare. The difficulty is not in measuring the relationship between climate and agricultural production at a given geographical point. We know very well how to go about it. Such relationship is both undisputable and important. Plant behavioral patterns even make it possible to forecast precisely what would be the production of this or that crop under a given weather condition. The real issue here is to know whether climate fluctuations affect very extensive areas up to the point of disrupting markets. Here, nothing is obvious.

---

[1] In mathematics, the word "chaos" has a precise meaning. In short, it refers to neither monotone nor periodical solutions of certain differential equations.
[2] See Leuthold and Wei (1998), Hommes *et al.* (1998), Hölzer and Precht (1993), Burton (1993), Lücke (1992), Chavas and Holt (1993).

The great drought of 1976 in France was a revealing anecdote. Everyone in this country remembers that disaster which led to the imposition of an exceptional tax referred to as "drought tax" to assist farmers affected by the drought. At that time, there was a French agricultural model under experimentation at *INRA*, which involved a "climate index" (Boussard, 1975). It was therefore tempting to use it to evaluate the damage with precision. Yet, the application of the 1976 parameters to the model in question resulted in almost no drop in production. The reason was that while meteorological indices were very bad in the west and mid-west of France, they were more than excellent elsewhere, thus offsetting the situation.

French agricultural production actually dropped in 1976, no doubt for other reasons: for two years, due to the increase in prices of oil products because of the Six-Day War, farmers took to reducing the quantities of fertilizers they applied, as indicated by statistics of the period. A decrease in the quantity of fertilizers applied to crops does not necessarily produce an immediate impact because there are stocks in the soil as well as other "buffer phenomena". However, in the long term, such practice would inevitably lead to a drop in production, and this is what happened in 1976.

This story is instructive in several respects. Firstly, it is comical that public opinion accepted without hesitation this drought story, whereas they could no doubt have readily blamed the Arabic emirs for another calamity, because the latter were responsible for the increase in oil prices[1]. In reality, in all cases of drastic decline in agricultural production, natural disasters are always easily accepted as the cause. This does not mean that such a reason is correct. Amartya Sen[2] has rightly demonstrated that famines, by contrast, are generally due to "non-technical" causes.

---

[1] One may also wonder whether they were really responsible…which would take us very far!
[2] See in particular Drèze and Sen (1989).

Climate abnormalities do not disrupt markets

The story also shows that in a relatively small country like France, meteorological events tend to offset each other. This means that the benefits to be expected from liberalization are probably limited since, in reality, they already have been generated by internal liberalization in most medium-sized countries.

Lastly, it reveals the difficulty involved in carrying out studies of this kind: the contingent and arguable nature of any average weather index for any large region. Indeed, when one takes a superficial look at one of the weather charts published by all daily newspapers, one has the impression that there are immense phenomena, covering vast expanses like the North Atlantic, or Siberia or the Sahel. Such an impression is not false at a given time; if "climate abnormalities" covered such vast expanses, they would obviously be the cause of serious market disruptions.

However, for a plant, what matters is not rain or temperature at a given moment[1]. Rather, it is what happens during a relatively more or less long period, varying according to the species and especially their stage of development, and which lasts for days and sometimes weeks. By contrast, weather phenomena are evaluated in hours, with none of them similar to the previous ones, or occurring in the same areas. Consequently, to prepare a weather index, it is necessary to evaluate the differences from the average calculated over variable periods. Ideally, it is necessary to elucidate the relationship between the instantaneous fluctuations observed and those that were observed in the preceding recent periods. Then the "correlation distance" – the distance beyond which statistical links between abnormalities become so weak that they are negligible - should be assessed. Only at this stage may one apply the law of large numbers. Lastly, one should check whether

---

[1] The only exception, from this standpoint, is frost. And yet, it affects the different plant species in different ways.

the areas involved for a given random factor, in this context, are vast enough for the corresponding production variations to disrupt markets... Such a program and such an accumulation of difficulties would daunt the enthusiasm of even a young doctorate student intending to tackle the problem!

The near lack of serious studies on the issue is no doubt due to these difficulties – and to the fact that the social system as a whole "does not want to know". There are some isolated studies. For instance, in a study that unfortunately has never been published, Martineu and Tissot (1993) attempted the adventure and investigated whether "drought" could well be the cause of the difficulties of the Sahel – as the idea is complacently spread, particularly by NGOs. The "correlation distance" to which we alluded above is about 100 kilometers. This excludes the notion of drought from Dakar to Djibouti, although without sidelining the possibility of perturbation in the local markets, where regions are much compartmentalized and transport is difficult. Besides, such a result, if it were also valid for France, would justify the low impact of the 1976 drought we referred to above.

Risk plays the role of negative technical progress

The issue may be addressed from a different angle. One may take a market whose collection zone covers a limited surface area and seek to determine whether the climate in the area is the cause of the fluctuations observed. This is what Richard Roll (1984) did with a lot of care in a study on orange juice in Florida. In fact, the citrus grove in that State is situated in an area that effectively covers a hundred square kilometers. In addition, orange juice is certainly one of the agricultural products subject to very limited State intervention. It was therefore tempting to use it for experimentation. The outcome was quite interesting: "The market price of concentrated orange juice is affected by the prevailing weather conditions, particularly low temperatures. However, there is a puzzle in the orange juice futures markets. Even though weather has the most obvious and significant influence on the orange crop, weather surprises explain only a small fraction of the observed variability in futures prices... There is much that is unknown about price volatility". If the weather plays such a limited role in the case of a product like this one where it is expected to play an obvious role, it means that it almost plays no role at all in the case of commodities available throughout the world. In addition, one may also think that "endogenous fluctuation" mechanisms, which we referred to above, are better placed to play the "missing link" role described by Roll.

Whatever the case, it is clear that such price fluctuations cannot, under any circumstance, be disregarded in the evaluation of the benefits of liberalization. In Chapter 1, we saw that the economic effects of trade were comparable to those of technical progress. Here, we see that by reducing production possibilities, risk very obviously plays the role of negative technical progress. If liberalization reduces risks, then everything would be fine and real benefit would be greater than expected in Chapter 1. However, if it turns out that it increases risks, the issue would then be to know if the disadvantages of increased risk could overcome the benefit of a better use of comparative advantages.

Do liberalization benefits offset risk disadvantages?

One may reasonably wonder whether the controversies surrounding the liberalization of agricultural trade that raged during the past centuries were not based on this argument. During each period, liberalization was no doubt tempting because the benefits of exploiting comparative advantages were obvious. But each attempt was obstructed by price volatility, which annihilated the benefits expected, and prompted a rapid backtracking. Is this not likely to be the case with the Doha negotiations[1]?

To answer this question, it is necessary to take two measures:

- try to calculate the order of magnitude of the benefits expected from liberalization, with the same objective as in Chapter 1, disregarding the objections that have just been raised. In recent years, many international agencies have tried this using "economic models" which we may call "standard";
- construct an alternative to the standard model that would take into account all the phenomena we have just discussed.

This is going to be the focus of the following chapters of this book.

---

[1]The Doha rounds centre on the liberalization of international trade (in particular, agriculture and services). Initiated in 2000, the negotiations resulted in a declaration after the fourth ministerial conference convened by the World Trade Organization in November 2001 in Qatar. The conference underscored the importance of the development objectives of countries of the South.

CHAPTER IV

Designing an Economic Model for International Trade

"We have to admit that on very many points, farmers are more knowledgeable on political economy than economists and policy-makers". Leon Walras

A global economy "model" is first and foremost a set of equations which link up a number of variables – income, prices, produced and consumed quantities. This set is constructed in such a way that the variables representing these entities behave within the model as the corresponding magnitudes would, where appropriate, in global economy. This is somehow a sketch, a "scale model" of the economy, which the experimenter may manipulate as a child would do with a video game, to see "what happens if..."[1]

Because global economy is clearly a quite complicated field, it is easy to imagine that the number of equations to be manipulated will be high. In fact, the model in question can only be a huge set of simultaneous equations which will require modern computer science to

---

[1] In fact, the reader must have understood that the idea is not to construct a kind of political utopia, a "model global economy" which would not suffer from the defects of real economy. It is a purely descriptive and hardly normative approach, aimed only at reproducing the functioning of a real economy. The word "model" here is used in the sense of "copy" and not "object to be copied".

be solved. Elaborating such a monster may appear like the sign of ridiculous pretense to accomplish a formidable and unrealistic task. How is it possible to even dream of such a complicated thing evidently void of any illusion of success?

However, as seen above, such models exist, and are even used by policy-makers to commit the future of Mankind. Of course, the policy-makers in question have no illusions on the accuracy of the results they are using. At best, they use orders of magnitude, and definitely approximate figures. Besides, even the initial basic data are hardly known: who can provide, with a 10% approximation, the amount of "global national income" for the year 2000? It is obvious that figures published in statistical yearbooks are very rough estimates. As such, trying to forecast them in ten years is a bit of a challenge, and may stand as an intellectual scam.

But, at the same time, it is certain that these models have at least the advantage of enabling the study of assumptions which are all interrelated: the production of each commodity is equal to its consumption (including inventory change), income is equal to expenditure, etc. Many absurdities frequent in customary discourse on futurology are thus avoided. By highlighting the ultimate consequences of a hypothesis or a set of hypotheses, this type of models may be used to buttress reasoning. They are thus considered by people who make use of them – at least, by the most serious of them – because, unfortunately, there is also a "misuse" of models, which consists in taking at face value the models that are in the line of what is wished…

However, in order to use the models "properly" as defined above, one has to somehow work extra hard, read between the lines and understand their design. Such is the subject of this chapter.

As a matter of fact, such a model is very simple in its principle[1]. There are producers and consumers in each country. Naturally, all that is produced must be consumed on the spot or exported or stored. All that is consumed must be produced on the spot or imported or taken from inventory (where it exists). Where the production and demand are known, writing out such relationships produces enough equations to determine equilibrium prices and stocks in a "simultaneous equations" system.

It is also necessary to determine the equations that enable to calculate productions, consumptions and incomes, which is not very obvious. Also, there is need to obtain the corresponding data to calibrate the model over a reference period. Lastly, it is necessary to define the dynamics of the model – the rules governing its behavior over time. These points will be treated successively. First of all, let's see how these "standard" models (those that aim to reflect the theory expounded in Chapter I) are constructed. An attempt will later be made to show how, with relatively minor change applied to these standard models, it is possible to bring into play the considerations developed in the preceding chapters on expectations and risk, which profoundly influences the results obtained and the ensuing practical conclusions.

Production

Producers maximize their profits, consumers their "utility"

In the production sub model, entrepreneurs are supposed to maximize profits. Under such conditions, they should balance the price at which they sell with the "marginal cost", the cost of the last item produced, or in mathematical terms, the derivative of the "cost function" with regard to the quantity produced. It links a total cost

---

[1] It is almost impossible here to provide a bibliography on the topic, given the numerous good references available. It is however worth mentioning the excellent abstract at the beginning of the book by Folmer et al. (1995), as well as Hertel and Tsigas (1997).

to each production level. Where the cost function is known, the equation written to that effect makes it possible (at least in principle) to determine the production level of each entrepreneur. But where does the cost function come from?

Ricardo's parable, which was presented in Chapter I (Box 1), gives an example of a particularly simple cost function: cost is directly proportional to production, whether in Portugal or in England, for fabric or wine. In fact, in each case, the cost expressed in hours of work, or, which is the same, in man-year, is "linearly" related to production and to the use of a single means of production, labor. In reality, things are not that simple, because there are several means of production or inputs: for wine production, in addition to labor, land, fertilizers, pesticides, etc. are also necessary. The same applies for fabric production. These various production means may however result in numberless combinations, with each being associated to a cost in a given price system. Of course, in each price system, entrepreneurs will strive to minimize costs, everything being equal. The choice of the production method and the production technique therefore depends generally on the prevailing price system. That is why the cost function must, at each production level, associate a cost which is the minimal cost, taking into consideration the price system – which itself depends on demand and supply. This is to say that, here again, we find ourselves in a complex system of simultaneous equations.

In practice, the cost function is associated to the "production function", which itself links each input set to an output, which is the quantity of products. This association is however not done haphazardly: everything is done to bring out the maximum quantity of output possible that can be obtained from each possible input set, in a way to take into account only the "efficient" productions, those that can not be increased without a corresponding increase of the quantity of at least one input, while the others remain unchanged... As a matter of fact, the "true" production functions are not known. Instead, they are attributed approximate algebraic expressions such as the "CES function" used in the numerical illustrations of this study (see Box 7).

> **BOX 7 THE CES FUNCTION**
>
> CES is the abbreviation for *constant elasticity of substitution*. In fact, constant elasticity of substitution functions are extremely convenient for economists who venture into the modeling of the major characteristics of production without really wanting to enter into details.
>
> We all know that a "production function" is a relation between an output (e.g. the quantity of wheat produced) and corresponding inputs (e.g. the number of hectares of land or hours of work, tractors, etc.). It is written as $y = f(x_1, x_2 \ldots x_n)$, where y is the output, and $x_1 \ldots x_n$ are the input quantities. Normally, such a relationship is very complicated: there are thresholds (1 tractor, 2 tractors, etc.) and nothing is *a priori* proportional. However, for purposes of a model like the one with which we are concerned here, one needs an f function easy to write, always defined, continuous and derivable, at least for positive x values. In addition, it should possess certain properties imposed by common sense, for instance: $f(0, 0 \ldots 0) = 0$ (where no input is provided, nothing is produced); $f'(x_i) > 0$ (the derivative of f with respect to x_i, the marginal productivity, must be positive); $f''(x_i) < 0$ (the more the quantity of an input, the less its marginal productivity).
>
> The CES function:
>
> $$f(x_1 \ldots x_n) = \alpha \, (\delta_1 \, x_1^{-\rho} + \ldots + \delta_n x_n^{-\rho})^{-1/\rho},$$
>
> where $\alpha, \delta_1 \ldots \delta_n$ and ρ are parameters that are expected to reflect the state of technology, suits many of these specifications.
>
> It is a "constant elasticity of substitution" function because the "input-output coefficients", that is $q_i = y/x_i$, are such that if producers maximize their profit, then $e_i = dq_i/q_i = \sigma dp_i/p_i$, where p_i is the price of the input i, with the other prices remaining constant. Coefficient σ is the same for all the inputs (which is a disadvantage and an over-simplification). It is a simple function of parameter σ which, in the case of only two inputs, represents the substitution elasticity between the two products (the relative variation of input 1 which should be accepted to reduce or increase by 1% the quantity of input 2 without changing production).
>
> It should be pointed out that, where the substitution elasticity is equal to 1, the CES function will be reduced to the popular "Cobb-Douglas" function : $y = \beta x_1^{\alpha 1} x_2^{\alpha 2} \ldots$

The hypothesis along which real production functions are CES functions makes it possible to easily and conveniently attribute a minimal cost and an optimum "input purchase program" to a given price system, for a given production level. This is what an entrepreneur normally does. One is thus at liberty to think – if the CES function is not a too gross approximation of input-output relations – that it enables to provide a representation of the latter's behavior, at least in a first approximation. Practically, this is done by writing out the entrepreneur's "first-order conditions", that is equations which express that the marginal productivity value of each input is equal to its price (Box 8).

Naturally, the process implies an equation for each input used in each production. Since, in addition, in order to bring out the comparative advantages, this should be written separately for each country, the number of simultaneous equations to be solved will very rapidly become astonishing. That is why it is necessary to limit the number of "products" (and of "countries" to be considered. In reality, it is clear that it is not the same thing to produce maize in a "dry farming" or an "irrigated farming" system, or with one variety of soil rather than another. But, considering the level of accuracy we are using, the issue is neither to distinguish between the various varieties of maize, nor even "maize" from "other cereals". We will therefore have a nomenclature of industry sectors which, even by focusing on agriculture, will comprise only about twenty products, such as "wheat", "other cereals", up to the "engineering industry".

All this is rather complicated and technical. We will not dwell on that more than necessary, while bearing in mind that, in the final analysis, both the production level and the techniques that will enable to obtain it depend heavily on the price system. We have seen above that prices were obtained through the equations which express equality between demand and supply. But how is demand determined?

> **BOX 8 "FIRST-ORDER CONDITIONS"**
>
> In order to look for the values of the variable which give an "extreme" value (minimum or maximum) to function $y = f(x)$ of a real variable, one has to write that the derivative $f'(x)$ of y with respect to x is nil: by solving the equation thus obtained, we obtain the values of x corresponding to these extreme values.
>
> The "first-order conditions" that we write in general equilibrium models, and in many other problems of mathematical economics, correspond to a generalization of this method in the case of several variables. Consider an enterprise which uses quantities $x_1, x_2 \ldots x_n$ of production factors 1, 2, n, to obtain the produced quantity y, knowing that the production function is given by:
>
> $$y = F(x_1, x_2 \ldots x_n) \qquad \ldots(1)$$
>
> and that the prices are p_y for the product, and $p_1, p_2 \ldots p_n$ for the factors. Its profit is given by $B = p_y y - (p_1 x_1 + p_2 x_2 + \ldots + p_n x_n)$. It will be hopefully maximal (it could be minimal!) if y and the x_is are such that for every i:
>
> $$p_y F'_i(x_1, x_2 \ldots x_n) = p_i \qquad \ldots(2)$$
>
> where $F'_i(x_1, x_2 \ldots x_n)$ represents the derivative with respect to x_i of $F(x_1, x_2 \ldots x_n)$ at the point considered.
>
> It should be noted that $F'_i(x_1, x_2 \ldots x_n)$ is nothing other than the "marginal physical product" of factor i, the additional quantity of product that one can obtain from an infinitesimal increase of the quantity of input i. Equation (2) therefore expresses that the marginal productivity value of factor i is equal to its price.
>
> What remains to be done in principle is barely to solve the system of equations (1) and (2) for every i to find the values of y and $x_1, x_2 \ldots x_n$ which maximize profit. However, where F and F' are not linear, some problems of mathematical calculation may arise.
>
> In the case of the CES function (Box 7), we have:
>
> $$F'_i = \delta_i (y/x_i)^{(1+\rho)},$$
>
> which is particularly simple, and which explains the popularity of this function among modelers.

Consumer Demand

With a given income, the consumer is a kind of entrepreneur who, instead of maximizing profit, maximizes a "utility function", subject to the condition that expenses do not exceed budget. With this assumption, and provided the utility function is known, the problem of the consumer is exactly the same as that of the entrepreneur. It is no doubt necessary to get an expression for the utility function, which is not easy. Just like for the production function, at this stage, more or less justified hypotheses are formulated on the analytical form of the function, and its parameters are assessed based on consumer surveys conducted at regular intervals by national survey institutes and statistical bodies. Further on in this book, the numerical illustrations are derived from the use of a consumption function referred to as "LES", described in Box 9. However, to use this function, it is necessary to find out the source of income.

BOX 9   THE LES FUNCTION AND CONSUMER BEHAVIOR

In order to represent consumer behavior in models, particularly to assess reactions to price or income changes, economists are used to admitting that the average household maximizes the "utility" obtained from a set of products, under the constraint that the overall expenditure value is equal to their budget. For that purpose, it is necessary to define a "utility function" which gives the degree of "well-fare" associated with any given consumption basket. Naturally, this is something quite hypothetical, because, clearly, no one has ever measured the level of satisfaction of anybody. However, from commonsense, one can obtain some information on such a function: it is certainly a function which increases with the quantities of goods consumed ("it is always better to have "more" than "less"), growth is no doubt diminishing (since there is satiety in everything), etc. We know, in addition, through introspection and through the statistical survey of household budgets, that some goods are more demanded than others under certain circumstances. For instance, food expenses increase with income, and less than proportionally to income.

From such observations, one can imagine algebraic functions with suitable properties to represent these behaviors. The linear expenditure system (LES) function is one of those. Let x_i be the quantity consumed of good i, c_i a minimum quantity of this good deemed "essential" and U the consumer's utility, we have the following equation:

$$Log(U) = \alpha_1 Log(x_1 - c_1) + \alpha_2 Log(x_2 - c_2) + \ldots$$
$$\alpha_n Log(x_n - c_n).$$

Coefficients α_i represent preferences and they must sum to unity.

When looking for quantities x_i which maximize U under the budgetary constraint, that is:

$$p_1 x_1 + p_2 x_2 + \ldots p_n x_n = R,$$

where $p_1, p_2 \ldots p_n$ represent prices and R the nominal income, writing the "first order condition for optimization" results in the relation:

$$\alpha_i = p_i (x_i - c_i)/(R - p_1 c_1 - p_2 c_2 - p_n c_n), i = 1 \ldots n,$$

which can be interpreted by saying that "the marginal share value of each product in the overall marginal consumption remains constant". Consumption increases in value are linear functions of the income, hence the name of this model. Similarly, the consumption of certain so-called "inferior" products increases less rapidly than

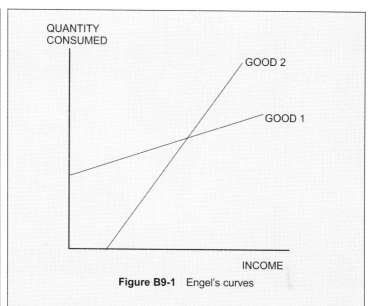

Figure B9-1 Engel's curves

income, whereas the consumption of other so-called "superior" goods increases faster. This makes it possible to incorporate the previous remarks on the low demand elasticity for agricultural products.

This type of functions allow to represent the most common behaviors, at least at first approximation, and in consideration of the situation which was used to estimate coefficients α_i. Here (figure B.1), two items ("1") and ("2") have been represented. For, good 1, consumption does not vary much with income. Good 2, on the contrary, is a luxury commodity, very sensitive to a change of conditions. Of course, the income has to be large enough for the consumption of good "2" to be positive in all the cases under study.

Income Determination

Income comes from the remuneration of production factors.

At this stage, we must have noticed the presence of two distinct sets of inputs in the model under construction. Some inputs are such that by their nature it is possible to produce them: for instance, copper sulfate used in vineyards may be produced by the chemical industry, as opposed to the soil. Such products are referred to as "recurrent inputs". They constitute a special and additional category of possible uses of the production (in addition to the final consumption, export and storage).

But for other inputs, referred to as "factors", the quantity is "fixed" and cannot be changed, at least at a given moment. A typical example is land or capital. Their prices are determined by scarcity. These prices correspond to the "marginal productivity" (the derivative of the quantity of output with respect to the available quantity of input) in value (that is the quantity multiplied by the price).

In all the models used to analyze the consequences of liberalization, one unavoidable hypothesis is that "households" (sets of ultimate consumers) have "rights" on the factors, or, to be more specific, on the associated income. This is obviously the case of holders of securities (like shares in companies): profits linked to the capital (fixed factor) are distributed among the shareholders proportionally to the rights they own, through the securities they hold. But, in this perspective, workers also have rights on their work, which is remunerated, in the same manner, on the basis of its marginal productivity value, and results in the payment of income depending on the contribution of each, and proportionally to that contribution (as a matter of fact, a part-time worker is paid half less than a full-time worker).

Taxation makes liberalism tolerable

Such an hypothesis faces many difficulties because it is clear that many incomes are not determined this way in the real world[1]. It is however acceptable *a priori* and it is often possible to manipulate it more or less successfully by adding *ad hoc* equations to the model. In the following pages, it will be admitted that such hypothesis is verified, under benefit of inventory.

Also, the adverse effects of this hypothesis are mitigated by another circumstance: in the model, there is

---

[1] Particularly, such a specification leaves no role to firms. It is as though each economic actor directly offers their production services (be it labor, capital, land or any other thing) to the market, which then uses them for the better. We all know however that market alone is unable to perform the tasks of coordination which are necessary to make the best of production factors. As such, it is necessary to pass through firms, i.e. "bureaucratic" bodies, in which coordination is carried out through hierarchy and not by the market. See Coase (1937) and the abundant literature on "new institutional economics".

in each State a "government" whose function is to levy taxes and redistribute them. These taxes may be levied at different stages; at consumption, during importation, exportation, on income, etc. They may be positive or negative (like agricultural subsidies). They are never "neutral" and always cause "distortions", in the sense that their effect on production or on consumption is never nil[1]. In fact, taxation cannot leave prices unchanged, and we have seen that the price system was the crux of the decision-making mechanism: it should therefore be expected that any change in the taxation system should have repercussions on all consumptions and all productions.

For instance, an agricultural subsidy, even when unrelated to production, will change land prices, and hence land revenue. As a consequence, certain category of households will have their incomes increased or lessened, which will consequently modify their consumption pattern, with repercussions on the demand for automobiles or antic jewels, etc.

The existence of taxation however produces a major effect on the philosophy of liberalism: it makes it tolerable. Liberalism has indeed been widely criticized because, through the action of marginal productivities and of the rights held on production factors, it implies a revenue distribution which may well not be in line with the wishes of citizens, and which may even be contradictory to ordinary justice. Taxation is aimed at correcting these effects, and may do so efficiently, on condition that it is meaningful and rightly directed. It is for the purpose of studying this type of issue that the early versions of the "general equilibrium models" were constructed a few decades ago, especially by Atkinson. Naturally, this presupposes that the various categories of households, "rich" or "poor", "urban" or" rural", etc., should be correctly identified in the model, which is not always the case.

---

[1] From this viewpoint, the scientific community agrees that whether directly linked to production or not, subsidies have an influence on the decisions of producers.

Whatever the case, this analytical framework allows to easily establish a correspondence between factor prices, the quantities used, as well as the table of rights linking the households and the factors on the one hand, and the incomes of various household categories on the other hand, provided one has the corresponding data.

International Trade

The Armington hypothesis, a very convenient gadget

A model like the one whose major principles have just been expounded may be constructed for a national economy, for instance "England", or a regional economy ("Cumberland"). It could also be envisaged at a "global" scale, by simply distinguishing as two separate products "wheat produced in the United States" and "wheat produced in France"[1]. The disadvantage of the process is that it multiplies the number of unknowns in the model. If the economy of a country is represented with about one hundred equations and unknowns (this is in fact very few and hence caricatured), we would require more than 300 equations and unknowns for three countries (because in addition to those of a country are imports and exports from one country to another), and much more than 1 000 (in fact about 30 000) with ten countries. This will likely complicate the numerical resolution of the model.

But there is another problem at this stage: "wheat produced in Europe", from the consumer's viewpoint, is not very different from "wheat produced in the United States". In fact, the pure international trade theory presupposes that both products are identical, and should therefore sell at the same price. If this were true, it would be impossible to observe (as all statistics show) wheat exports from the United States to Europe and exports from Europe to the United States. Either wheat is more

---

[1] In this case, in addition to the equations which express national markets equilibrium are those which report on the equality between the sum of exports and the sum of imports of each product, as well as balance of payment equilibrium.

expensive to produce in the United States, and the flow is from Europe to this country, or vice versa. And were the costs identical on either side of the Atlantic, there would be no trade at all, since, in addition to the cost of the product there are transport charges, which entrepreneurs could not be so stupid as to bear.

One way of solving this difficulty is by considering that the "foreign" product is slightly different from the "local" product[1]. It is attributed a "production function" (which may besides include transport costs) and it is presumed that the product supplied to the consumer is the result of a special production operation, in which the only inputs are the "local" product and the "imported" product. A special entrepreneur (called "Armington", from the name of the author who seems to be the first to have proposed this modeling artifice) uses these inputs to manufacture the final consumer product by minimizing costs, taking into consideration local and international prices. This process increases the number of equations and unknowns (all the more as imports are distinguished by their origin), but is efficient and widely accepted.

Data Sources

National accounting and social account matrices

It is however necessary to find the corresponding data. Such data come from national accounting, in the form of "social account matrices". A social account matrix is a table of figures whose columns represent the expenditures of the various industry sectors or categories of economic actors, while the lines represent incomes.

---

[1] This is however very true: for instance, it is quite natural that a baker in Detroit (Michigan) gets flour supplies from a miller in Toronto (Ontario) rather than at El Paso (Texas). However, such a transaction will lead to importation of wheat from Canada to the US, while the purchase at Laredo would not be registered in international statistics. In effect, wheat from Canada has a particular "quality" here in the eyes of the Detroit buyer as compared to the average US flour... which may have the same distinctive quality for some Mexican buyers (it is obviously the same reasoning from Cuidad Juarez to El Paso in relation to Acapulco).

There are as many lines as columns and the total of each line is equal to the corresponding column (table I).

Thanks to the efforts of specialized United Nations agencies, social account matrices are available for almost all countries of the world. However, they are not of the same quality everywhere, and are not always comparable. As a matter of fact, it is necessary not only to have data in all countries of the world, but such data should be coherent: the sum of exports of a particular good in all countries should be equal to the sum of imports, etc. Unfortunately, there is nothing as difficult as obtaining such sets of coherent data, because customs vary from one country to another and what statisticians consider important is not necessarily important to others. As such, definitions hardly correspond, balances do not concern the same things and there is no compatibility between various elements.

For the numerical illustration presented here, we drew much inspiration from the pioneer work carried out at Purdue University by a group of researchers led by Thomas Hertel (see Box 10). They developed the GTAP database which, today, is an unavoidable reference for any research of this nature.

BOX 10   THE GTAP DATABASE

GTAP is the abbreviation for *Global Trade Analysis Program*. It is a gigantic project conducted by Thomas Hertel and his team of Purdue University. From the onset, it was an approach very similar to that presented at the beginning of this essay, to set up a global general equilibrium model, without embellishing it with accessory consideration relating to risk and expectations. GTAP was thus one of the "traditional" models for which we are seeking other forms of construction.

However, the GTAP project quickly developed into two separate projects: a standard global general equilibrium model, as mentioned above, and the corresponding data. The latter is normally very difficult to collect, since it entails constructing worldwide a coherent national accounting system, based on an accounting system not yet implemented by most national statistical offices.

Tribute should be paid to the tenacity and sense of diplomacy of Professor Hertel for successfully standing up to both activities. We no doubt have reservations on the model (but, it was the best at the time it was designed) and we know, as well as Professor Hertel himself, that some of the data concerned are questionable, and roughly measured. All the same, this exceptional piece of work is today a must for whoever intends to go through such studies.

Table 1 Example of SAM – this table traces the activity of the major "aggregates" of Polish economy in 1991. Expenditures are entered in columns, while incomes appear in lines. For instance, the remuneration for "non-farm work labor", which represents 304.68 units of account t(UA), is allocated as follows: 1.27 UA to "farm households" (result of part-time farming), 240.10 UA to "non-farm households" and 63.32 to "social welfare". By reading the table per line, we see that the same total of 304.68 UA is derived from "non-farm households" income for: 18.42 UA from food industries, for 134.81 UA from "other industries", and for 145.34 UA from "services". The table reveals that the simple definition of the titles of lines and columns reflects by itself a rather in-depth analysis of the problems facing the country. Besides, this example was chosen because of its relevance to the study of rural development.

An example of social account matrix: Poland in 1991 (trillions of zlotys)
(Source: Orlowski, World Bank, quoted by Boussard and Christensen, 1996)

	Goods and services consump.	Labor factors Farm	Labor factors Non-farm	Capital factors Farm	Capital factors Non-farm	Households Farm	Households Others	Current accounts Firms	Current accounts Adm.	Current accounts Social welfare
Goods and services consumption						43.44	436.95			
Farm labor										
Non-farm labor										
Farm capital										
Non-farm capital										
Farm households		34.22	1.27	6.93	0.00				0.78	8.82
Non-farm households			240.01	0.00	152.98				10.92	124.05
Firms (other than agric)					168.5					
Administration					13.9	0.33	3.35	109.19		
Taxes and duties										
Social welfare			63.32		0.00	0.54	5.42		63.60	
Capital accounts – agriculture						8.44				
Capital accounts – others							84.85	59.31	-26.74	
Agriculture and fisheries	36.78									
Food and agric industries	96.21									
Other industries	114.17									
Services	233.23								167.97	
Rest of the world	0.00								30.30	
Total	480.39	34.22	304.68	6.93	335.37	52.75	530.56	168.5	246.83	132.87

Table Contd.

Table Contd.

	Capital accounts		Activities				Rest of the world		Total
	Agric	Other sectors	Agric & fisheries	Food & agric indust.	Other indust.	Services	EEC	Others	
Goods and services consumption									
Farm labor			34.22	18.42	134.81	151.44			480.39
Non-farm labor				18.42	134.81	151.44			304.68
Farm capital			6.93						6.93
Non-farm capital				23.72	166.31	145.34			335.37
Farm households							0.55	0.18	52.75
Non-farm households							1.89	0.63	530.56
Firms (other than agric)									168.50
Administration			0.00	6.45	44.58	32.57	7.21	2.40	219.99
Taxes and duties			0.76	2.56	21.36	2.16			26.84
Social welfare									132.87
Capital accounts – agriculture			4.75						13.18
Capital accounts – others				1.40	12.99	13.56	16.01	5.37	168.60
Agriculture and fisheries			35.32	37.52	1.71	10.17	4.12	6.24	131.87
Food and agric industries	0.12	1.70	1.23	20.36	3.97	24.31	9.66	6.01	163.67
Other industries	5.58	82.35	25.01	15.13	50.33	191.81	63.00	59.85	607.32
Services	5.72	84.54	18.52	21.08	29.22	240.04	20.70	4.73	825.76
Rest of the world			5.05	17.02	142.03	14.35	0.00	0.00	208.75
Total	13.18	168.60	131.87	163.67	607.32	825.76	123.24	85.50	

Factor endowment	Agric & fisheries	Food & agric indust.	Other indust.	Services	Total
Capital	757.67	89.28	898.65	1 735.03	3 391.35
Labor (thousands)	3 963.80	406.90	3 390.90	6 500.90	13 855.60
Land (thousands of ha)	14 636.00	x	x	x	14 636.00

Thanks to the GTAP database, it is possible to develop the model, without much effort, based on data provided in table II, for thirteen regions and seventeen sectors.

Table II Regions, sectors and factors in the CIRAD ID^3 model

13 Regions	17 Sectors	5 Production factors
Europe	Wheat	Skilled labor
United States	Other cereals	Unskilled labor
Australia, New Zealand	Cattle rearing	Land
Central and Eastern Europe Countries (CEEC)	Other animal production	Natural resources
MERCOSUR countries	Milk	Capital
Other Latin America countries	Oil crops	
Developed Asian countries	Sugar	
Southeast Asian countries	Other crops	
South Asian countries	Silviculture	
China	Agro-food industries	
Sub-Saharan Africa	Meat	
North Africa, Middle-East	Milk	
Rest of the world	Sugar	
	Others	
	Timber industries	
	Other industries	
	Energy sector	
	Services	

A simplified version of the model was also prepared with only three regions, "the United States", Europe" and the "rest of the world", without any changes in the sector and factor nomenclatures. The corresponding results, in what follows, are presented as "results of the three-region model".

Computers for economic equations

In some respects, this broadly presented global economy model is an extraordinary monument of simplicity and genius. It is the result of two hundred years of research in economics, since the earliest ideas could be traced back to the French king Louis XV's physician, François Quesnay (1694-1774), founder of the physiocrats school, with key improvements from authors like Adam Smith (1723-1790), Léon Walras (1800-1866) or Wassily Léontief (1906-1999). Whatever the genius of the economists who designed it, it would have remained a theoretical curiosity without the data that accompany it, and above all, without computers, which are indispensable to solve gigantic simultaneous equation systems. These instruments are relatively new. About ten years ago, it would have been unthinkable to develop a model like the one whose results we are going to give, without having very costly and very difficult to program[1] specialized computers. Today, with the aid of software like Gams and its solvers[2], the solution can be obtained in a few minutes on an ordinary office computer, which is not at all designed for that purpose, since modern computers are designed to process images and sounds rather than for mathematical calculation.

Dynamics: The Role of Time in the Model

As described above, there is however one missing key dimension in the model, that of time. As a matter of fact, economy evolves with time, and decisions are taken at intervals, one very often conditioning the other. Under

---

[1] Even today, such a model does not operate like word-processing. In solving ID[3], the authors of this book were faced with severe mathematical calculations problems. More than once, they did not find a solution for the equilibriums they were looking for (which, in principle, were feasible). Computational algorithms are based on the idea of seeking the equilibrium through a system of "intelligent" trial and error process starting from a point which, in principle, is arbitrary. But if the starting point is too far from the equilibrium sought for, it is practically impossible to find it.

[2] Gams - General algebraic modeling system - is a programming language specially adapted to the processing of input/output data, designed in the 1970s by the World Bank. It comprises solvers and mathematical software specialized in optimization, making it a reference tool in that domain. It is however far from being the only one of its type.

such conditions, one has to admit the rather unrealistic nature of the "Walrassian model" (from the name of Léon Walras, whose role in the development of this model in the 1850s has just been underscored) described above, in which everything is simultaneous and where everything instantaneously reacts to everything.

How dynamics can contradict statics.

This shortcoming is especially unfortunate for a model of international trade. In effect, after Ricardo, nobody can dispute the advantages linked to trade between economic entities of different nature. However, throughout the 19th and 20th centuries, many authors questioned the advantages of "free trade". However, they never tried to brush aside Ricardo, but most often based their work on the dynamic characteristics of the economic game. The most representative among them is no doubt Friedrich List (1789-1846)[1], who explains that although trade is an excellent thing, excessive liberalism should not prevent new industries from developing. If we want an impartial measure of the advantages or disadvantages of liberalism, we absolutely have to introduce such dynamic considerations in the model. But how?

In reality, as we will see later in this book, the problem is fundamental, and is the source of most of the misunderstandings between the "liberals" and "protectionists". The former have in mind a harmonious and foreseeable world: they cannot imagine that trade could have disadvantages. The latter, more pessimistic, find expectations and risk errors everywhere: they are thus very prudent.

---

[1] A German by origin, he stayed briefly in Paris (where he even considered publishing his book in French), then he emigrated to the United States and came back to Hamburg as the United States Consul. He finally published his *"National system of Political Economy"* in German. See the fascinating preface to the French translation by E. Todd (List, 1849).

To translate such considerations into a model, no doubt the ideal is to start from the most optimistic and also the simplest (these are not its only advantages!) "liberal" model. With this model that we refer to as "standard", and which is currently promoted by international organizations like the World Bank, the Organization for Economic Cooperation and Development (OECD) or the International Food Policy Research Institute (IFPRI), it will be possible to assess the benefits of liberalization under the most favorable assumptions. Then, it will be possible to gradually introduce modifications in the model to make it more realistic. This will thus allow not only to obtain more accurate and better evaluations for policy-makers, but also to measure the effect of simplifying hypotheses which have been introduced in the standard model to make it more flexible and easier to calculate, but which might not reflect reality.

We should therefore start by explaining how time is taken into consideration in the standard model. Given that time plays therein only a minor role, this would be quite easy. However, even for this simple version of real economy, there are still several variants.

The stationary version of the dynamic model

The "pure" way to view time in the so-called "neoclassical" model is to consider that an object identified at two different moments is in fact two different objects. As such, a model of this type will not talk of "wheat produced in the US", but of "wheat produced in the US, 2003 harvest". This would be a different object from "wheat produced in the US, 2002 harvest". Conceptually, this does not change anything in the model: instead of the equation "wheat market equilibrium", there will be two equations, "2002 market equilibrium" and "2003 market equilibrium", as so on and so forth. This modeling trick makes it possible to easily and most naturally introduce considerations on savings and investment which were obviously lacking in the so-called "Walrassian model" expounded above. Along that reasoning, a one hundred-year old oak would be obviously different from one that is 101 years old. But, in addition, a 100-year old oak in 2003 "produces" in 2004 one 101-year old oak. (In strict logic, not really one, since 100-year old oaks are subject to a certain mortality rate, in such a way that, probably, and pending in-depth consultation of an oak mortality table, in terms of expectation, one should no doubt expect a rate of about 0.99 or 0.98). The same reasoning can be applied to just any equipment, which makes it possible to write equations to express the market equilibrium of capital goods, and consequently of savings. This approach is in accordance with the old "Austrian" theory of investment[1]: we save today in order to have the possibility of consuming more tomorrow thanks to the "productive roundabout" investment. The interest rate is a market price that matches supply with demand for savings, i.e. the exchange of immediate consumption for greater future consumption.

This fascinating conception of time makes it possible to make a projection into the future in a perfectly mechanistic manner. Is it realistic? One may want to ponder over this question, and this will be done later on. What is clear is that a model based on this principle poses serious problems of mathematical calculation: we have seen that a roughly accurate representation of an "annual" global economy required about 30 000 variables and equations. If we have to multiply this figure by the number of years, we would see that, to solve the same model within a time-frame of ten years using the just specified framework, one would require 300 000 equations. It is not realistic, at least in the present state of computational sciences.

---

[1] From Böhm-Bawerck (1851-1914), famous Viennese Statesman and economist.

For that reason, the authors of the international models mentioned here have generally adopted a solution presented as more modest. It consists in representing the economy for just one year. During this year, as seen above, all decisions are taken simultaneously, in such a way that, for instance, an upward trend in wheat price will immediately produce an increase in supply, which will tend to bring back the system to equilibrium. However, savings is an exception to this rule. It is considered simply as a constant fraction of income, determined without taking into account either the interest rate or any other adjustment variable[1]. Such savings is immediately transformed into capital goods, thereby generating demand for such goods, and thus contributing to the general market equilibrium. This of course is an "average" capital good comprising a mixture of nuclear plants, combine-harvesters, etc.

Furthermore, in the most standard version of the standard model, this unique capital good is added to the unique capital good stock which constitutes the "capital" factor of the model. This stock is however reduced for depreciation for wear and tear, to calculate the capital stock of the following year. This model is thus transformed year in year out in a "recursive" manner, with the current year capital stock derived from the results of last year. It is worth pointing out that the stock of this average unique capital good is distributed "in the best possible manner" among the productive activities at the time of calculating the equilibrium of each year: There is no obstacle using a combine-harvester to produce electricity, since it cannot be distinguished from a power plant.

To avoid what is all the same an absurdity, many models use another conception of capital. Instead of one stock of undifferentiated capital, there are as many

---

[1] This artifice is often presented as a "Keynesian" version of the models. If that were true, it would be a caricature of the Keynesian theory, all the same clearly richer. In any case, it is a convenient solution.

capital goods as of products. Such a process generates many complications. First, the so-called "recurrence" equation (installed capital in year n = installed capital in year n – 1, + investment, – depreciation) must be written for each industry sector, instead of being unique for the entire economy. This would not be difficult to perform if one did not have to know which fraction of the overall investment should be allocated for each sector. Moreover, the demand generated by building an automobile factory is not exactly similar to that of an investment in pharmacy. In most cases, for want of *ad hoc* data, this second difficulty is avoided: reflection continues on an average "capital good", without taking into account the sector to which it is allocated. On the contrary, the former difficulty, that of knowing to which sector the new capital should be allocated, is unavoidable, and many solutions have been sought for it.

All these solutions are based on the observation of the marginal productivity of the capital. In effect, the software used for solving general equilibrium models always provides such information, which is indispensable to calculate the "price" of the capital, and thus the remuneration of the households which hold the rights thereof. From these data, many models use a "price elasticity" of capital supply: if, for instance, the capital of the "agricultural" sector increased by 1% last year, then the share of agricultural investments in overall investments will increase by x%. This method is no doubt convenient, but it does not reflect the true problem of the investor. The latter diversifies his portfolio depending on the gains he expects (reflected by "last year's price of the capital"), but also the risks he runs.

As we have seen, whatever the solution chosen, this "recursive model" process is far from satisfactory to the most intransigent of economists. To them, it is nothing but a second-best solution. The "real" model to them seems to be the great multiperiod model, whereas the recursive model is nothing but an ersatz. At the same time, one cannot understand how economic actors would be much more skilful than the most skilful economists to forecast forthcoming equilibrium prices and quantities in the long-term. That is why it is no doubt quite intelligent to be contented with the recursive model, but also, there is every reason to believe that it is more realistic that the other.

The Standard Model

In any case, many people are contented with the standard model. That is why some of the results which will be presented in the following chapters were obtained with the latter, which consequently comprises:

- A momentary equilibrium of production and consumption, without any risk and without expectations, with CES functions (Box 7) for production and LES functions (Box 10) for consumption;
- a rather unsophisticated recursive dynamics, which however permits, to a certain extent, expectations and risk considerations for the distribution of savings among the various industry sectors;
- a break-down of household data into the "rich" and the "poor" – with the "rich" representing the upper half of the consumer population classified according to income ranks, and the "poor" the lower half;
- fixed production factors per sector, with, however, the possibility of changing the use of some of them – for instance, unskilled labor may be moved from one farm activity to another, likewise land.

Model adopted by international organizations, in line with Ricardo (18th century)

These specifications make it a rather "high-quality" model among the common models, but without any particular originality. Moreover, for the CES and LES functions, we repeated the values of the parameters mentioned by Thomas Hertel (1995) and which had been used in the model by this author[1]. Not that such values were "better than others" or estimated more accurately, but simply to avoid speculations on the fact that our results were obtained using "unorthodox" parameters which we could be required to justify. Lastly, all the data were collected from the GTAP databases compiled by Hertel and his collaborators (see Box 10).

Under such conditions, it would have been surprising for our results to be different from those already published a long time ago by the GTAP team. As a matter of fact, they are not. We will see in the following chapters that these results are totally in conformity with those that have been presented on several occasions during international meetings[2] and which play an important role in persuading the public that "liberalization is good".

But it is also obvious that these models are in line with the analysis presented some two centuries ago by Ricardo. They absolutely do not take into account the ensuing remarks and reflections. So, would it not be possible to construct a more original model, which would take into account the phenomena observed in the last two centuries? Such is the subject of the next chapter.

---

[1] In reality, he uses a constant difference of elasticity (CDE) function. It is possible to switch from one function to another.

[2] The "GTAP community" meets each year in a grand colloquium which permits everyone to present their findings. During the 2002 meeting held in Taiwan, about 300 models of this type were presented, with, generally, as only real specificity, that of bearing not on the entire international trade, but on the effects of liberalization on one or other province or State. The practical results were in general similar as they all recommended liberalization, so much so that it appears useless to belabor the point here.

CHAPTER V

How can Theory and History be Introduced in a Standard Model?

From a standard equilibrium model to a more realistic disequilibrium model

How can the model described in Chapter IV be modified to include considerations on the behavior of the economic man and their macroeconomic consequences discussed in Chapter II? To achieve this, it is necessary to analyze the role of risk and expectations in the formation of equilibriums. Indeed, expectation errors and the associated risk impinge with more or less intensity on behaviors represented in such a model.

First, there are producers' expectation errors. The price at which I, the producer, will actually sell my products will generally be different from my expected price. Through cobweb phenomena (Chapter II), the consequences of these errors are numerous and induce chaotic price series.

There is also the problem of capital formation. In Chapter II, we have seen how difficult it is to admit that capital can be distributed among the sectors in an "optimal" manner. The problem is the same for labor. "Fixed factors" are the cause of this situation. Naturally, risk considerations still play a major role here.

Lastly, there are issues of foreign exchange risks, international (or national) capital flows, and more generally, interactions of "financial" circles with the "real" economy. Here again, with "perfect" markets, as the great liberal economist, Jean-Baptiste Say, puts it: "there is nothing as indifferent as money[1]". This is why the "standard" model does not consider it. Nonetheless, we've known for a long time – this is the core of the Keynesian message – that, from the perspective of the imperfect functioning of markets, the way the economic man behaves towards different financial assets can have serious consequences on the functioning of the "real" economy. This aspect should be taken into account when designing models.

Expectation Errors

It is very possible to take into account considerations regarding imperfect information, as analyzed in Chapters III and IV, and to adapt the standard model accordingly. This is what was done in the ID^3 model developed by CIRAD, which we are going to describe before presenting the results in the next chapter.

The first thing to do is to re-examine the dynamics of the model. In the standard model, production and consumption are simultaneous. Like in the ideal supply and demand equilibrium model, prices alert producers about possible disequilibria "in real time" and the latter react instantly by increasing or decreasing their production.

However, we have seen that things do not occur that way. We also saw that, far from being secondary, the phenomena resulting from the likelihood of short-term supply-demand distortion can become a determining factor – and particularly in agriculture, where demands are stiff.

---

[1] "Il n'y a rien de si indifférent que la monnaie".

In the model described in Chapter II, it is easy to create a supply-demand lag. We can decide that the production of year t + 1 is decided in year t, based on price expectations formed in year t for year t + 1. Entrepreneurs therefore carry out their cost-benefit analyses using these "expected" prices, exactly as they would have done with equilibrium prices, and resolve their "first-order conditions" in the same manner. In this way, they create demand for current inputs and factors. This will increase final demand for year t, and immediately modify the price equilibrium. However, the available supply in year t would have been determined in year t − 1, based on expectations of year t − 1 for year t. In this way, the actual prices at which goods would have been sold in year t will differ from the ("expected") prices used by entrepreneurs to determine their production. Entrepreneurs derive positive or negative "profits[1]", differences between "what had been expected" and "what occurred". After all, this is something that each person can observe everyday. At the same time, such a mechanism is likely to produce cobweb phenomena, which were obviously

---

[1] It should be recalled here that the CES functions used in the standard (as well as in the modified) model exclude *a priori* the notion of profit. Due to their mathematical characteristics (these functions are "homogeneous and of degree 1), the production value is automatically equal to cost. Of course, from the accounting viewpoint, a profit does exist nevertheless. It is the difference between the value of production and the value of inputs which are not owned by the entrepreneur, and must be bought on markets. Therefore, this accounting profit corresponds to the reward for the entrepreneur having committed private property in production. However, this is not a profit in the economic sense of the word because this remuneration of the private property of the entrepreneur is computed "at market value", exactly as if the entrepreneur had bought them from other people. There is no difference between the full production cost, and the total value product.

excluded from the basic hypothesis of the standard model, the immediate adjustment of supply to demand[1].

At this point, it is necessary to describe how "expectations" are formed. We could, as Ezekiel presumed (see Chapter II), consider the price of last year as the best estimator for the price of this year. Should we, however, ignore the fact that the decision maker is not so naive as not to "sort through" this type of data? To take account of this, several years ago, Nerlove (1958) devised an "adaptive expectations" system, in which the expected price for next year is the price of the current year corrected by the observed difference between the expected and the actual price for that year. In this way, it can be proven (Nerlove *et al.*, 1995) that expectations are made of a weighted average of all previous prices, with weights decreasing with time (memories of distant past play a smaller role than those of recent periods).

However, whenever there exist differences between "what was expected" and "what occurred", risk occurs. Perhaps an entrepreneur will be lucky enough to perform better than expected, but performing worse cannot be excluded. No actual decision can ignore that. The equations representing the behavior of entrepreneurs in the model should therefore be modified. These modifications are quite simple: the expected price should

---

[1] Of course, the hypothesis of a one-year lag between supply and demand is highly questionable. Actually, there are many lags. Many years separate the design of a new car from its arrival on the market. In the agricultural sector, although decisions relating to cultivation of crops like wheat or rice are taken yearly, several years are required for a tree to bear fruits. Conversely, stock adjustments in trade take much shorter periods and seldom exceed a few months, if not a few weeks. An annual adjustment is therefore a rough estimate. Two reasons, at least, justify such a rule. First, it is better to integrate an imperfect and rough lag than to have no lag at all. Second, this is an average for highly "aggregated quantities". It is therefore possible to admit, though this is debatable, that the lag between fruits and vegetables is an average between that of apples, which takes several years, and that of tomatoes, which is three months.

be replaced by its "certainty equivalent", which does not significantly change the equations.

One must admit that the notion of certainty equivalent itself is subtle (Box 2). For a price or any other random quantity, and a given decision maker, the certainty equivalent is the value which, if known with certainty, would induce the same decision as the random variable under examination[1]. This is a very complicated notion. In practice, however, in most models, the certainty equivalent is calculated like a weighted sum of the actuarial expectation and variance of the variable, which is rather arbitrary but better than nothing.

Naturally, this requires that the price variance should be known. Unfortunately, it is not. In effect, it must also be "expected". The best estimator is unquestionably the squared deviation between the expected price and the price quote. This is what has been retained in the ID^3 model. An adaptive expectations formula could also have been considered for the variance, like for the average.

At this stage, two difficulties are still to be solved to "close" the "annual" version of the model. Firstly, the mechanisms that have just been described imply that there are profits. In fact, profits do not exist in standard models. As cost usually equals price in these models, profits always remain nil. In our case, they reappear as a consequence of risk taking, along the most orthodox economic theory. They are sometimes positive and sometimes negative, although the existence of risk premiums ensures that they are globally positive in the long run. However, they must be transferred to income. We will see later what would have been the correct method to do this, from a financial standpoint. For the

---

[1] For instance, imagine a lottery ticket with one over two chances of winning $1000 tomorrow. If I decide that I am ready to buy it for $400 and no more, then the certainty equivalent of this random prospect for me is $400. Of course, it normally differs from the mean, which, in this case, is $500.

version of the model used here, to keep things simple, profits have been paid to capital holders proportionately to the capital held.

The second problem stems from the lag between investments and savings. In standard models, investments and savings are simultaneous. As a result, the demand for capital goods in the form of securities always equates savings. In physical quantities, the latter reduces the demand for consumer goods and increases the demand for capital goods; but equilibriums in all markets are easy to establish.

The situation is not the same here, as savings are made in cash. The corresponding demand for capital goods arises during the next period only, when sectors in which these savings are invested are known, as will be seen later on. Conversely, the demand for goods of all kinds is equal to the consumption resulting from the income of the current year plus the demand for capital goods resulting from savings decisions taken during the previous period. That, however, does not significantly modify the general structure of the model.

Under such conditions, table III provides a general picture of the differences between the standard and the

Table III Comparison between the standard model and the ID^3 model: the "annual equilibrium" section

Equations of the standard equilibrium model	Equations of the disequilibrium model
1. Production of the current year + imports = consumption + investments + exports	1. Production of the previous year + imports = consumption + investments + exports
2. Price = marginal productivity	2. Expected prices − risk premium = marginal productivity
3. Price = marginal utilities	3. Price = marginal utilities
4. Price = production cost	4. Expected prices − risk premium = production cost
5. Income = expenditure	5. Income = expenditure
6. Income = factor price x stocks held	6. Income = (factor price + profits) x stocks held
7. Imports value = Exports value	7. Imports value = Exports value

ID^3 models, at least for what concerns the "annual equilibrium" section. These changes, as we can see, are rather minor.

Capital Accumulation

We have seen above the problems arising from the introduction of time — and therefore capital — into this type of model. Physical capital does not actually move from one sector to another. For instance, one cannot produce electricity with a combine-harvester. Consequently, once savings have been invested in a combine-harvester, one cannot possibly use it to produce electricity.

The simplest general equilibrium models do not consider time and have only one "capital" sector. They thus implicitly assume that such a substitution is possible. We will not waste time to comment on the corresponding outcomes, which cannot be very serious. Others, however, rightly consider the amount of capital invested in various sectors of industry as distinct objects. Thus, the scarcity of capital in an expanding industry cannot be compensated by the availability of excess capacities in a declining or slow-growing sector. Nevertheless, as we saw in Chapter II, there is a problem relating to the rules governing the management of this multiple capital stock that can be compared to an equipment pool. What does one do with savings obtained at the end of a given year? In which sector should it be invested? For a specialist of economic modeling, the most natural way to address the issue is to draw on "elasticity" as defined in Box 5 (Chapter II) for what concerns demand.

The process consists in assuming that if the profitability of an investment for a given year increased by x%, then the proportion of total available savings which will be invested in this sector will increase by y% (a few precautions in choosing parameters are necessary to ensure that the algebraic sum of the increases and decreases is nil). This way of modeling the behavior of investors is certainly preferable to the hypothesis along which they will always miraculously discover the most promising investments, with automatic adjustments ensuring equal return on capital in all sectors. Thus, allocating fresh capital to sectors in this way is "better than nothing", and guarantees consistent results. Nonetheless, many questions remain unanswered.

That is why other models — like the one presented here — use the so-called "Markowitz" sub-model (see Box 11) to distribute new capital among sectors. This model is generally regarded as a good description of investors' behavior. It introduces a fundamental concern, namely risk, in the calculation.

In fact, with this sub-model, risk plays a key role in the diversification of investments between sectors. Without risk, total savings would be invested in the "most profitable" sector. However, as soon as risks are envisaged, it becomes clear that the "most profitable" sector is also the most volatile. Thus, part of the savings will be invested in less risky sectors. This is how private investors behave. The consequences of such behaviors are felt at the macroeconomic level.

BOX 11 THE MARKOWITZ MODEL

The holder of a securities portfolio naturally seeks to make the most of it. However, maximizing the expected profit generated by the portfolio, under no other constraint than the total sum to be invested, automatically yields a strange result: the totality of the investor's wealth must be invested into one single, "most profitable" security, which presents the best prospect for profit. Yet, all financial analysts know that at a given time, the "most profitable" security is also the "riskiest" one — which next month can be worth either a fortune or nothing at all. Furthermore, none of them would advise their customers to purchase a large amount of such a security in a portfolio, even if it is often recommended to buy "a few".

To consider this aspect, a long time ago, Markowitz introduced the notion of "efficient portfolio" as a mixture of "high potential" but risky and low-yield but gilt-edged securities — with a dosage that depends on the security holder's "risk aversion". Technically, that is done by maximizing a function that is a linear combination of profit expectation and variance.

> To express the above ideas with more precision, let the standard problem (which does not consider risk) be stated as: maximize $F = \Sigma(\bar{p}_i - c_i)$ under constraint $\Sigma c_i x_i = M$,
>
> where: x_i represents the number of certificates of share i that appear in the efficient portfolio; p_i and c_i correspond to the resale and purchase price, respectively, of share i; the mathematical expectation (the average) of p_i is represented by $\bar{p}_i$, M is the sum to be invested in the portfolio. Then, the optimal solution to this problem is simple: $x_i = 0$ everywhere, except for i which corresponds to the maximum value of $p_i - c_i$, for which $x_i = M/c_i$.
>
> By contrast, Markowitz's problem is therefore to maximize:
>
> $$U = \sum_i (\bar{p}_i - c_i)x_i - A \sum_i \sum_j x_i x_j s_{ij},$$
>
> under the same constraint as above: $\Sigma c_i x_i = M$, with, in addition to the preceding notations:
>
> s_{ij}, covariance between p_i and p_j;
>
> A, risk aversion coefficient of the portfolio holder.
>
> The resulting portfolio is much more diversified

The model described here is constructed according to these principles. The other modifications indicated on table IV follow from this. (We will not dwell on these models any longer, and refer the readers to detailed descriptions of the model and its variants for more information[1].)

Table IV The dynamics in the standard and in the ID³ models

Dynamics in the standard model (exists in some versions)	Dynamics in the ID³ model Recurrence equations
1. Savings (t) = % income (t)	1. Savings (t) = % income (t)
2. Investment in value (t) = savings (t)	2. Investment in value (t) = savings (t)
3. Capital (t) = Capital (t − 1) − depreciation (t − 1) + investment (t − 1,j)	3. Capital (t,j) = Capital (t − 1,j) − depreciation (t − 1) + investment (t − 1) (a) Profitability (t) = function [year price (t -1)] (b) Investment (t, j) = function [profitability (t − 1, j), volatility (t − 1, j)]
4. New physical capital allocated to the most profitable sectors for equilibrium	

[1] For example, see Boussard et al., 2002.

This way, the "general equilibrium" model is transformed into a "disequilibrium model". That does not mean that markets are unbalanced: indeed, *ex post*, supply always equates demand everywhere, which should prevent us from talking of "disequilibrium". But the quantities supplied are determined on the basis of expectations which do not guarantee equilibrium *ex ante*. Indeed, since, each year, the supplied quantities are fixed exogenously from last year production, equilibrium is based solely on price variations. The latter are therefore more important than in the usual case where quantities vary concurrently with prices.

Such a concept is in no way extraordinary in economics: the idea dates back at least to the end of the 19th century, when the Swedish Knud Wicksell used it to explain economic crises (Chapter II). The extraordinary thing today is that professional economists still do not consider as self-explanatory an idea so commonplace and so obviously consistent with everyday experience.

In fact, "standard" general equilibrium models highlight the flaws in any kind of agricultural policy by comparing their outcomes with an ideal, but probably unworkable system. Conversely, by addressing the problem of market volatility, the ID^3 model considered here makes it possible to evaluate the benefits of the agricultural policies designed to reduce volatility (as seen in the preceding chapter). The aim of these policies is not to deviate clumsily from a sort of economic paradise that would be the perfect competitive situation, but to resolve real problems relating to the actual functioning of markets, which are not behaving exactly as they are expected to do in a theoretical Eden.

The Role of Firms, Banks and Financial Markets

The considerations above have been incorporated into the ID^3 model. Those described hereunder are still being designed. They are nonetheless essential.

Until now, the role of firms and institutions has not been addressed. Of course, as soon as risk is considered, there is room for State interventions, especially those aiming at decreasing price volatility. In a model, this represents a progress over the standard Walrasian approach. But firms and banks do not yet play a specific role, from the perspective of Coase (Chapter IV).

In a famous article published in 1937 (whose profound originality and great relevance is being understood only now), Coase explained that if markets were perfect, there would be no need for firms. Owners of production factors would sell them directly on the market, which would ensure no skill is unused and that no production prospect is lost. Coase regrets that the market does not function well enough for this to happen. The difficulty in coordinating the actions of men is such that institutions like firms are necessary. However, decisions in firms are fundamentally "bureaucratic", or "hierarchical". They have nothing to do with the market. Certainly, the market guides them, since the firm, as an organization, seeks to maximize profit or growth through the market. But it does not mean that there is a market for everything done within firms.

Standard general equilibrium models do not take this aspect of economic life into account. Indeed, they are made in such a way that they imitate the functioning of the perfect Coase's market in which holders of rights over factors lend them to "production" without any intermediary. In "social account matrices", firms consist only in dummy accounts merely intended to show the way in which proprietary rights are distributed between households. So far, the ID^3 model, as it has just been described, obeys this rule.

Nevertheless, there is a genuine need to show firms as such in general equilibrium models. On which basis could a model be specified to this end?

First, it is necessary to assign them an economic role. In keeping with a tradition traced back at least to Cantillon[1], the most natural solution consists in assigning them a role in risk management. Firms can therefore be presented as institutions that distribute dividends to their owners. Titles of ownership of firms would be exchanged in a stock market and the purchase of a security would give rise to a share of the profits (or losses) generated by the firm.

Profits themselves would be the difference between the revenue of firms — the proceeds of their sales — and factor costs — labor, capital or other resource — they would buy on the factors market with the money contributed by their shareholders or borrowed at a fixed rate. Firms would use these factors to manufacture one or more products, each of them requiring certain factors defined by a production function. The overall products of each firm should not require the use of factors that exceed those that the firm owns or buys. Of course, firms would be risk averse, which explains their profits (plus factor incomes) through the existence of risk premiums.

Households themselves would sell their labor on the labor market and buy titles of ownership from firms using their savings. They might also buy "resources" like land or greenhouse gas emission rights and lend them to firms. They might obviously buy from foreign and national firms in as much as this is authorized by national regulations defined by economic policies. That would

---

[1] Richard Cantillon (1680-1734), an economist and demographer, author of *Essay on the nature of commerce in general*, is considered as one of the pioneers of economics. He had been impressed by the failure of the "Law system" – an attempt for replacing gold by paper money in the early 18th century France, which ended with the ruin of most speculators. From that unfortunate experience, he knew what a risk meant.

pave the way to the analysis of international capital movements, which is currently not taken into account by economic models.

Such modifications of the standard model would not be very difficult to implement. Yet the results presented below do not take account of these considerations. The differences between the standard and the ID^3 model are based only on the first series of discrepancies presented at the beginning of this chapter.

Illustration of Differences

We will now focus on detailed results. As a conclusion to this dry chapter, the scope of differences between standard models and disequilibrium models will be illustrated by the comparative results of the two types of models in a simplified but very illustrative version.

During the preparatory work for the results presented in this chapter for the standard model and which will be presented in the next chapter for the disequilibrium model, the two types of models were operated comparatively, in their "three-region" version. In fact, the standard model in this experiment is not exactly the standard model, because the financial module is the same in the two models and incorporates risk. Besides, the disequilibrium model is perhaps less imbalanced than it is in subsequent versions because price expectations are "constant", such that the extent of price variations is significantly mitigated. Lastly, the data is not that of GTAP98, but of a preliminary version, GTAP95.

With regard to original wishes, these restrictions have a real, though unexpected advantage: it is possible to operate the model for fifty years, almost like the chaotic cobwebs presented in Chapter II, whereas in the subsequent, much more realistic version, it is unusual for the model to function for more than ten years. Beyond, due to numerical problems themselves related to very high price variability, it is usually rare to find a solution with the "final" model (that for which we give detailed results). This is certainly very annoying and frustrating to the authors of the model. But, that undoubtedly testifies to its greater realism.

Indeed, in their design, and at least to define the "reference situation" (which will be used as a basis to assess policy assumptions whose effects are to be tested), these models imply the continuation of "current policies" throughout the simulation period. Yet, although a simplified model can support "current policies" for a long time, recent history clearly shows that these policies are seldom enforceable over a very long period of time without modification[1]. Therefore, it is rather a sign of quality that this model, by not finding any more solutions within the "current" institutional framework, suggests that these policies are no longer viable after a few years.

In spite of this situation, and to have a very long-term vision of things, it seemed useful to operate the simplified model, which does not stop, over a long period. The results are presented in figures 9, 10 and 11.

Figure 9 highlights the difference in global GDP (the sum of GDPs of each country or region) between the two hypotheses, with and without liberalization. In the reference solution, without liberalization, policies in force at the time of data collection, that is 1996, are retained. In the version with liberalization, all obstacles to trade, whatever their origin, are removed. This difference was assessed using each of the two variants of the model, first the standard model (Chapter IV), and then, the ID^3 model. We see clearly that the results are not the same. The standard model (grey line) indicates a

---

[1] The example of the US Fair Act of 1996 immediately comes to mind. Initially envisaged to last five years and inspired by a real concern to achieve liberalization which was to be pursued thereafter, it did not last more than three years before necessitating the adoption of a multitude of costly emergency measures eventually incorporated into the new 2002 agricultural act.

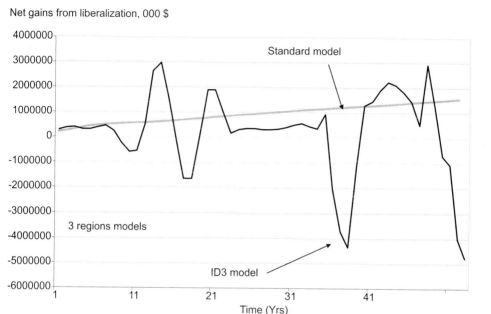

Figure 9 Discordant evaluations of liberalization-related gains and losses by 2 three-region models: the standard model and the ID³ model

relatively low evaluation — about one thousand billion dollars, that is less than two hundred dollars per capita —, but constantly and regularly increasing advantages of liberalization. The ID³ model (black line) shows significant changes in gains and loses. Generally, over a period of about fifty years, the total losses are much higher than gains.

The study of the distribution of these gains confirms this impression. Figure 10 displays the results of the standard model. These are deviations (expressed as a percentage of the reference solution, without liberalization) of the scenario "with liberalization" with respect to the "continuation of current policies" reference scenario. Here, differences are calculated for the incomes of the "poor" and for those of the "rich" in the three regions.

Such results are striking in view of their "politically correct" aspect: the main winners of liberalization are the

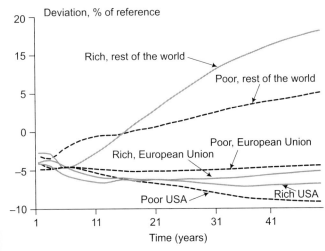

Figure 10 Three regions standard model: Percentage trend of the incomes of the poor and the rich with liberalization, as compared to the reference situation without liberalization.

"rich of the rest of the world". (This includes underdeveloped countries, which make up the majority of the "rest of the world"!) Admittedly, it would have been preferable for the "poor" to be the major beneficiaries, but one should not ask for too much, inasmuch as the "poor" of the rest of the world are not fairing badly either in this result. They rank second in the list of beneficiaries. The losers are in the United States, a few of the rich, and slightly more of the poor. For the European Union, the differences are negligible, and rather tend toward losses, with the rich losing a little more than the poor. On the whole, nothing better could have been expected, considering after all that the United States, like the European Union, has the means to pay compensation to their losers.

However, one is disappointed after examining Figure 11, which indicates the results obtained with the ID^3 disequilibrium model for the same liberalization scenario.

Indisputably, these results are quite different from the previous ones. In fact, they are catastrophic, except

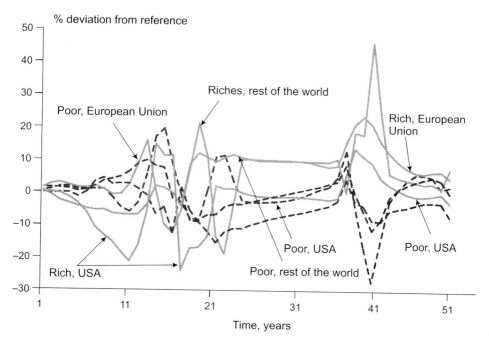

Figure 11 Three-region ID³ model. Percentage trend of the incomes of the poor and the rich with liberalization, as compared to the reference situation without liberalization.

during the very first years, where short-lived gains are observed, and in line with the ones highlighted in the previous figure (the "rest of the world poor" making some gains, which are nevertheless low).

After a few years (just time enough for more sophisticated models to realize that they are no longer viable), almost everybody lose, except at the end of the period. The rest of the world poor lose more than the others do. Negative gaps are significant. From this viewpoint, it should be noted that the scales on the two graphs are not comparable. On the first one, gains and losses crawl around +/− 5%, while on the second one, they vary between 10 and 20%.

Naturally, all these results with the second model reflect the negative effects of the malfunctioning of markets. Because entrepreneurs receive risk premiums and because markets are unstable[1], average prices raise far beyond technical production costs. Although this is translated into benefits for the wealthiest farmers, it also causes disasters for consumers, especially poor consumers.

That's why, when designing agricultural policies, it is important to consider endogenous fluctuations instead of limiting oneself to the apparent benefits of liberalization, which in fact are unachievable. Can such results, which are obtained from a simplified and truncated model, be generalized? This is what we are now going to find out.

---

[1] It should be noted that there is not the least "random number generator" in this model. Hence, the fluctuations are exclusively endogenous, as explained in Chapter II. In fact, the whole aspect of "climate fluctuations" was eliminated from this study to focus entirely on the problem of endogenous fluctuations.

CHAPTER VI

A Choice of Results

What can we learn from a more realistic model?

The results presented in the preceding chapter are final: if world markets operated conveniently, liberalization would generate a modest but real and significant increase in the overall income. However, although we take into account the good reasons we have for believing that markets do not operate as explained in elementary textbooks, it should be admitted that these results are far too optimistic. In view of the detrimental consequences of expectation errors on the price fluctuations they generate, it could instead be asserted that liberalization would deteriorate rather than improve the global economic situation (which of course does not mean that the current situation is the best possible). We could somehow stop at that.

The fact remains that the model used for this demonstration is "simplistic", with only three regions: the European Union, the United States and the rest of the world. Could these results have been confirmed by a more detailed model, where the world would be represented by a wider variety of countries? (Or rather regions, since it has to be admitted that the great number of countries does not make it possible to get down to such level of details). We will try to experiment this, and seize the opportunity to explore part of the richness of these results (only a part, since the mass of data extracted from a single "run" of the model is enormous).

There is real difficulty, however. In its multiple-country version, the ID^3 model – the imperfect information model which, as we have seen, provides a better representation of reality – is difficult to run over a time long enough to enable the construction of graphs like those that were presented in the previous chapter. This, as we noted, is a consequence of the fact that it better represents reality. As a matter of fact, never has an economic policy been applied for long without any change. In most cases, a change in policy follows the discovery that continuation of the existing policy is simply impossible because it leads to insurmountable obstacles[1]. It is therefore not surprising that a "realistic" model could not operate over a very long time under the hypothesis of "continuing the current policy".

However, by chance, in at least one very specific (and perhaps unrealistic) situation, it has been possible to get "quite long" series with the ID^3 model in its two versions, with and without liberalization. Detailed results will be presented below. Before that, nevertheless, we need to compare them with those of the three-region model presented above and check on a more realistic model the validity of the conclusions regarding the negative consequences of liberalization.

Does the Detailed Model Confirm the Results of the Three-Region Model?

At first sight, this is not the case, as shown by Figure 12, which was constructed in the same way as Figure 9

---

[1] In Europe, for instance, the establishment of milk quotas in the late 1970s was decided under pressure from the necessity (which had become evident for everyone at the time - there were even plans for a Greenland glacier storage, where cold was expected to be free!) of halting the accumulation of butter and milk powder stocks.

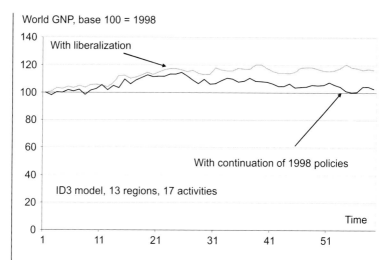

Figure 12 Gross world product trend with and without liberalization, in the thirteen-region ID^3 model

(Chapter V): the situation "with liberalization" is noticeably and constantly better than the situation "without liberalization". However, this conclusion is misleading. The gross domestic product gain obtained with global liberalization is profitable to Europe only! If

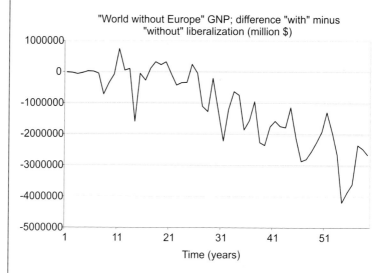

Figure 13 Gross domestic product trend for all the regions but Europe, with liberalization, compared to the reference situation, without liberalization, in the three-region ID^3 model

we remove this region from the overall, this gives us Graph 13, which very much resembles Figure 9 (but for the worse!).

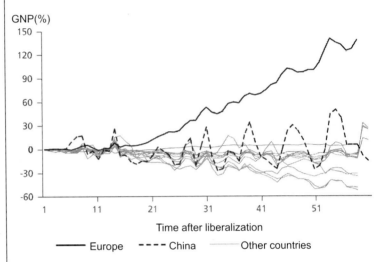

Figure 14 Percentage, on a regional basis, of the gain or loss of gross domestic product with liberalization, compared with the reference scenario, without liberalization, in the ID^3 model

Two issues are raised here: why and how does Europe succeed in benefiting alone from liberalization? What explains that the other countries do not reap almost any benefit?

Figure 14 confirms this assertion. It gives essentially the same results, but for each region. In order to limit the number of curves, instead of presenting "with" and "without" liberalization curves one after the other, the percentage GDP "with" – GDP "without"/GDP "without" - is displayed for each region. It comes out clear that with this scenario, liberalization produces a big winner, the European Union, whereas almost all the other countries are losers. Those who gain therein do so on a temporary basis only, and soon plummet anew.

Why may Europe Benefit from Liberalization?

The answer to question one is that, Europe would not benefit from liberalization simply because it would surely not continue the 1998 common agricultural policy model for fifty years. Yet, it is precisely this policy - far more than liberalization - that is devastating in the "reference" scenario (the scenario "without liberalization". The evidence is that liberalization with the ID^3 model is unfavorable to the United States, simply because the American policy at the time (which was a reference for that region) depended on instruments that were different from those of the Common Agricultural Policy (CAP). It is worth understanding why the 1998 CAP model led to such deplorable results, and we will now attempt to find the causes of this failure.

The database provided by GTAP (the global social accounting matrix used for developing the model, as explained in Chapter IV (Box 10), corresponds to the 1998 global situation. At that time, the United States strongly liberalized their agriculture, while the EEC stuck to the 1992 CAP reform system, with yet relatively high intervention prices (although significantly lower that those that prevailed before the 1992 reform). At the same time, in application of the Marrakech Agreements, there was a significant drop in customs duties, thereby opening the European market to imports from other countries. This, in principle at least, paved the way for third-country producers to sell in the Community, while there was no ready market for domestic production.

However, the intervention price mechanism is developed in such a way that it compels the EC to store excess quantities when supply is large and buyers ready to pass down the public intervention are not found fast enough. In the case under study, this is what happens, as seen on Figure 15.

An examination of this graph reveals that even when prices are determined at a rather low level, within a few years, the intervention price policy results in an extension of stocks beyond all limits. This is but a logical consequence of a prior expansion of the world agricultural production, which increases faster than

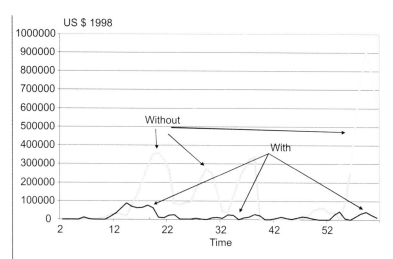

Figure 15 Thirteen-region ID^3 model : Trends for stocks in Europe with and without liberalization.

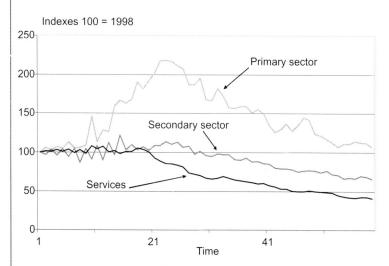

Figure 16 Thirteen-region ID^3 model : Evolution of the added values of the three sectors in Europe, without liberalization (assuming continuation of the current 1998 policies)

production in other sectors (Figure 16), taking advantage of the EC policy, which, at its own expense, stabilizes markets for everybody. However, such storage has to be financed, and this impinges on the overall savings capacity, which is thus reduced.

That is why one can notice on Figure 16 that, with such policies, all European productions tend to decrease over time (and not only agricultural productions, but also the industry and services!). Agricultural production growth would have been a mere fickle. In all, it would have simply choked the rest of the economy.

This situation contrasts with what can be observed on Figure 17, which repeats the elements of Figure 16, in a liberalization hypothesis. It is clear at this time, that the three sectors grow as expected: the agricultural sector remains sensibly constant, while industry and especially services grow most, as economic growth allows households to spend more by concentrating their increasing demand on the most desired goods.

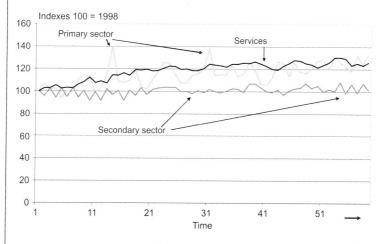

Figure 17 Thirteen-region ID^3 model: Evolution of the added values of the three sectors in Europe with liberalization.

Obviously, a scenario like the one represented in Figure 16 (with the stocks increases of figure 15) is not conceivable in reality. From the early signs of abnormal stock swells, one would have seen political circles panicking and imagining solutions that would be efficient at least in the short term. From this viewpoint, graph 18, which represents the real stock trends in the

EEC since 1970, is very instructive. It can be noticed that all the major reforms of the CAP coincided with stocks peaks, to the extent that one may even wonder if stocks increase was not the main reason or, at least, the trigger of the reforms...

In effect, neither the 1992 version of the CAP nor the American version, the 1996 Farm bill, lasted very long. As early as 1999, the United States came back on their liberalization decision, while the 1992 Mc Sharry reform in Europe was seriously reviewed, among other reasons because of stocks increase. It is therefore impossible to compare the model with reality as experimental rigor would require. Here is one of the shortcomings of economics, not to authorize this type of investigation. Conversely, the advantage of the model is that it makes it possible to propound the logic of a policy up to its extreme consequences over a long period of time.

From this standpoint, it is worth noting that the diagnosis which has just been presented could not have been made with the standard model alone, as shown by Figure 19, which gives the same results as Figures 16 and 17, but obtained with the standard model instead of the ID^3 model. Here, it is clear that the solution with liberalization is "better" than the solution without liberalization, but there is almost no dynamics (the growth curves are very regular and there are no qualitative differences from one year to another), to the point that it is difficult to trace the origin of the problem apart from the shady idea that the comparative advantages are better exploited. In reality, the assessed advantage of liberalization – or rather, of the change of economic regulation method, because we are now going to see that liberalization is not the problem – is considerably underestimated with the standard method, which does not really allow analyzing the root causes of the problem.

In this particular case, we can clearly see the disaster caused on the European economy by a permanent price

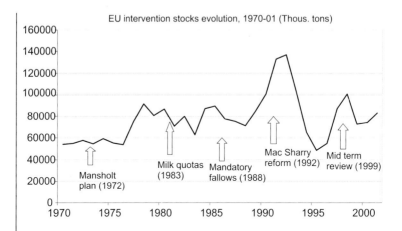

Figure 18 Historical trend of stocks in the European Union, from 1970 to 2001, in billions of tons (Source: Paris, Ministry of Agriculture, Daf.)

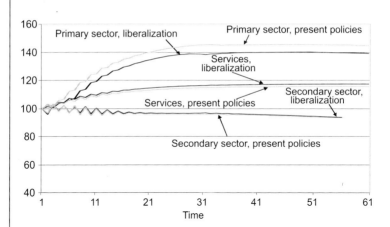

Figure 19 Standard thirteen-region model: Evolution of Value Added in Europe for three main sectors with and without liberalization.

guarantee without real limitation in quantity. The fact that liberalization frees the European Union from this concern is thus a positive point in favor of liberal solutions. It would be necessary to examine its consequences for the rest of the world.

Since the results are many and abundant, we need to choose. We will start with the case of a particularly poor and underprivileged region, sub-Saharan Africa. We will then continue with the case of a relatively rich region, one of the leaders of the Cairns Group, Australia-New Zealand.

Effects of Liberalization in Sub-Saharan Africa

Contrary to what was going on in Europe, the results of the standard model and the ID^3 model in sub-Saharan Africa, as illustrated in figure 20, are clearly different. The standard model shows that agricultural liberalization has almost no impact on Africa. This result contradicts both the early assessments by the standard general equilibrium models (which asserted that Africa would benefit from the opening of markets) and the more recent standard models, which hold that agricultural commodities price increase would further impoverish the urban populations. However, with the ID^3 model, although not very clear at the beginning of the period (there are ups and downs), the result is globally negative.

Figure 21 gives the reason for this result, at least for one example: the installed agricultural capital which, after a small increase which lasted for a short time,

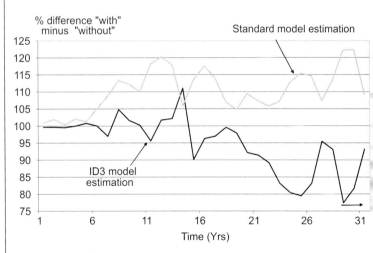

Figure 20 Percentage of gross national product gains and losses in sub-Saharan Africa with liberalization, compared with the reference situation, without liberalization, in the thirteen-region ID^3 and standard models

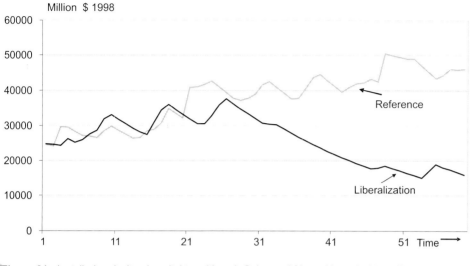

Figure 21 Installed agricultural capital trend in sub-Saharan Africa, with and without liberalization, in the thirteen-region ID^3 model

What repercussions in Africa for the rich and the poor?

sharply dropped with liberalization, as compared with the "without liberalization" reference situation. This is what happens in almost all sectors. Yet, capital is no doubt essential to produce in Africa: as a result of the cheap labor force and capital shortage, capital goods have a considerably large productivity, such that one does not need to greatly reduce the quantity thereof to cause very significant production drops. But why would the available quantity of capital diminish in Africa as a result of liberalization?

It is at this juncture that the risk considerations mentioned above come into play. In the ID^3 model, capital renewal is subject to a dual limitation because of the price risk: on the one hand, producers are careful and refuse to commit themselves in hazardous speculations; on the other hand, investors are also careful and hold back liquidities when the expected return on capital does not cover the risks. That is why the more prices are volatile, the more investments are scarce. But here, due to liberalization, prices become more and more volatile, as seen on graph 22.

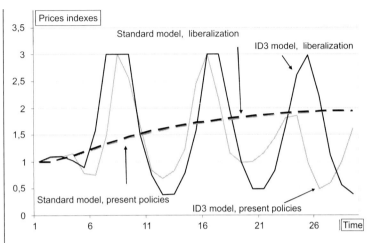

Figure 22 Thirteen-region standard and ID^3 models, with and without liberalization : Trends for cereals prices in sub-Saharan Africa

This graph shows trends in the prices of "cereals excluding wheat" in the ID^3 model and the standard model with and without liberalization. (Given that the quantity of wheat in sub-Saharan Africa is small, these are in fact prices for all cereals.) The two curves that represent the results of the standard model are so close that they can hardly be distinguished. In fact, if markets function properly, liberalization would not significantly change the prices of cereals in Africa (which further puts into perspective ongoing discussions on whether the "export subsidies" of countries of the North can discourage African producers or, whether the protection enjoyed by farmers of the North prevents Africans from gaining access to profitable markets).

However, it is necessary to further examine what occurs in imperfect markets. In this case, the graph shows firstly that even in the reference situation, prices are much more volatile than envisaged in the standard scenario. Figure 23, which shows some real producer prices in a quite typical African market, gives the impression that the ID^3 model is closer to reality.

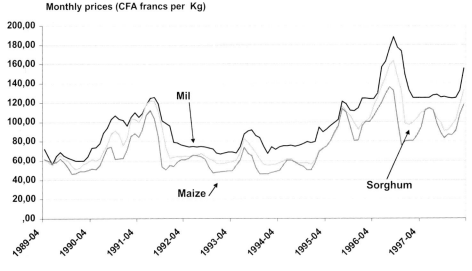

Figure 23 Monthly prices of cereals observed in an African market for maize, sorghum and millet in CFA francs per kilo from 1989 to 1997. (Sources: Market Information System, Club du Sahel. The data presented here was recorded in Sikasso, Mali.)

In fact, prices in Africa were already free in 1998 due to structural adjustment policies. From this perspective, the liberalization described here is not expected to change much. But the standard model, which does not take this type of mechanism into account, cannot explain it. This is what accounts for the difference between the results of the standard model and those of the ID^3 model.

Meanwhile, Figure 22 also shows that liberalization leads to changes in pricing policies: In the ID^3 model, the prices of the liberalization scenario are more volatile than those of the reference scenario. Thus, the (relative) supply and market control policies of countries of the North – in this case, the European Union – finally result in stabilizing (though to a small extent) the markets of developing countries and are rather advantageous to the latter.

Do these results correspond to the reality? Here again, it is difficult to compare the model to reality. However, it should be noted that the foregoing analysis is rather well

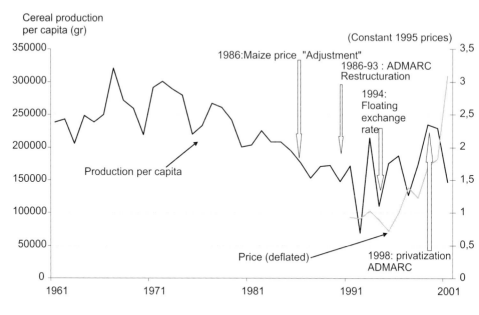

Figure 24 Per capita cereal production and value in constant prices (1995) in Malawi. (Source: FAOSTAT for production and current prices; The World Bank for the gross national product deflator, Charman, 2004, for the chronology of policies.
ADMARC: Agricultural Development and Marketing Corporation (price support agency).

verified in many cases in Africa. For instance, Figure 24 shows the long-term trend of cereal production in Malawi. This country had been the "maize granary" of Southern Africa for a long time. In the sixties and seventies, whereas prices were more or less guaranteed by a semi-public company, the country was exporting maize, and was time and again in a position to provide assistance to neighboring countries in the grip of famine.

From 1985, international supervisory bodies decided to dismantle this organization which, in actual fact, was costly for the national budget (in exactly the same manner as for the European Union, as we have just seen). The outcome was the extreme volatility of cereal production from 1990, as shown in Figure 24.

Production was characterized by tremendous (and useless) peaks followed by sharp falls, which caused severe famine leading to pressing food demands from the country to the World Food Program. On the whole, the general trend in food production (in line with population growth) has not changed. Yet, the resulting hardship – especially that of the poor – has increased significantly: one only has to look at the price curve to understand what the poor had to put up with!

This is explained by Graph 25, which represents the trend of the "utility[1]" of the rich and poor, in the ID^3 model in sub-Saharan Africa, subsequent to liberalization. This needs no comment.

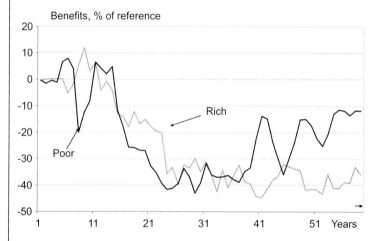

Figure 25 Thirteen-region ID^3 model : Percentage of gain or loss of household utility in Sub-Saharan Africa with liberalization, compared with the reference situation, without liberalization.

Naturally, these results differ greatly from those of Figure 26 obtained using the standard model. With the latter, the deviations to be ascribed to liberalization are much more modest in both directions Furthermore, they are relatively positive in the long term, even though this optimism is not unambiguous in the case of Africa.

By predicting an increase in the prices of agricultural products, the standard model concludes that African farmers would be better off than before, while poor city

---

[1]See Box 9 (Chapter IV).

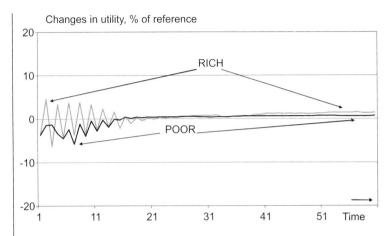

Figure 26 Thirteen region standard model : Percentage of gain or loss of household utility with liberalization in sub-Saharan Africa, compared with the reference situation, without liberalization.

dwellers would be less so. One of the most serious objections to liberalization by even the most liberal economists is based on this point: if there are losers in the liberalization game, winners should pay compensation to losers with part of their gains. Such a scenario, obviously, has little chances to obtain. Yet, in a dynamic model, and in the long term, optimists hold that after carrying out necessary but fairly costly adaptations, everyone gains, as shown figure 26. The previous figure 25 puts this argument into perspective.

To Conclude

Even when underestimated, instability is always present.

It would be very tedious to repeat the same detailed analysis for all countries. The case of sub-Saharan Africa, however exemplary, is by no means isolated, as shown by Figures 27 and 28, which themselves represent only one sample of similar results.

Figure 27 concerns two "emerging" areas, India-Sri Lanka on the one hand, and China on the other hand. Figure 28 concerns two "rich" regions: developed Asia (Korea and Japan) on the one hand, and Australia-New

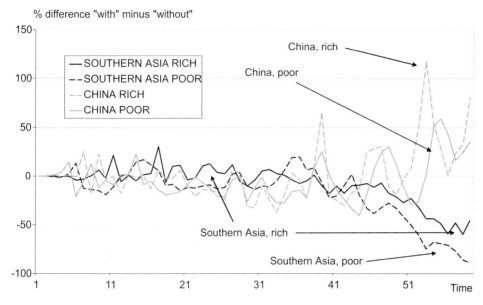

Figure 27 Percentage of gain or loss of household utility for two emerging regions: China and South Asia (mainly India and Sri Lanka), with liberalization, compared with the reference situation, without liberalization, in the thirteen-region ID^3 model.

Zealand on the other. Both figures were obtained from the series provided by the ID^3 model.

Admittedly, liberalization, according to the ID^3 model, is generally not of much use to households, and even less for the poor than the rich, although no rule seems to be very clear on the subject. It should be noted that Australia and New Zealand are hardly better off than India and Sri Lanka. However, China makes more or less giant strides, sometimes for the rich, sometimes for the poor. In any case, it is clear that even if (always temporary) winners do exist, the extent of the losses of losers would likely cause profound social unrest, which would soon make this type of policy unbearable.

Once more, the standard model gives much more optimistic results: undoubtedly, gains are lesser than those observed, where they are available, with the ID^3 model. However, they are much more regular and generalized, except perhaps, as we have seen, for the

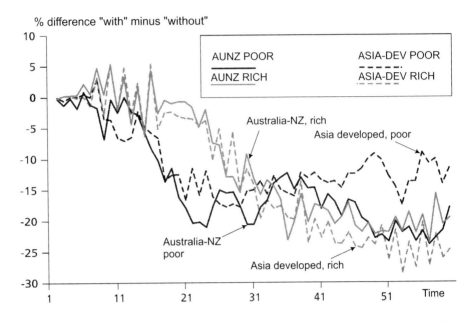

Figure 28 Thirteen-region ID3 model: Percentage of gain and loss of household utility for two wealthy regions: Australia-New Zealand and developed Asia (mainly Japan, Taiwan and South Korea), with liberalization, compared with the reference situation, without liberalization.

poor in Africa. One can thus understand why the World Trade Organization embarked on its liberalization program. However, it should be admitted that the results of the standard model are, at any rate, fragile.

This is not to say that those of the ID^3 model are less so: this model has many defects likely to ruin its conclusions. Monetary phenomena have been ignored. International trade transactions are carried out in a kind of "single currency" which does not suffer from any potential errors on the part of speculators on one or other currency. Now, it is quite clear that the instability of world prices is very much aggravated by exchange rate fluctuations.

Migratory and demographic phenomena are also ignored: although labor moves freely on the most profitable agricultural speculations, agricultural labor is stuck in the agricultural sector, just as industry and service-sector workforces are stuck in these sectors. Of course, this is an oversimplification. In fact, labor migrates from one sector to the other, even though these migrations are impeded by lack of qualification. (A farmer cannot be transformed easily into a medical doctor). Consumers also migrate from rural areas to cities and from poor countries to rich countries.

Private storage is another phenomenon which is ignored, though it plays an important role, at least in agriculture. In principle, the speculator who, in periods of abundance, stores agricultural products to resell them at a high price in periods of scarcity contributes to market stabilization. To that end, it is possible to take advantage of the numerous opportunities offered by "derivatives", (options or futures), which, in principle, enable everybody to accurately assess their risk as they think best. Lastly, private storage helps resolving a serious problem in the domain of food: harvests are done annually, whereas needs are felt on a daily basis.

All the foregoing considerations have been left out in the ID^3 model. It was to make ID^3 as comparable as possible to the standard model, which has exactly the same defects. Nevertheless, one may obviously wonder what the consequences of this negligence would be.

With regard to the foreign exchange market, there is no doubt that its introduction would have contributed to increase instability. On the contrary, the latter would normally have been reduced by taking storage into account and using derivatives markets. However, one should not entertain much hope in this regard. It is indeed very likely that the consideration of storage could have modified the physiognomy of price curves presented in the figures above, and perhaps even made them more akin to "real[1]" curves. However, that would certainly not have made it possible to eliminate these

---

[1] In any case, this is what was obtained by Di Costanzo (2001), whose theoretical series in the presence of speculative storage resemble real curves much more than our own.

fluctuations. On the contrary, authors like Deaton and Laroque (1992) demonstrate that due to limited storage capacities, storage can lead to price hikes, which are quite prejudicial to consumers when stocks drop to zero[1], which is inevitable.

As regards migrations and population growth, it is necessary to point out that we would not have dared to disregard them if the other models had not done so. These are massive phenomena that are likely to interfere seriously with world food balance in the future. Undoubtedly, such disturbances would have political consequences. It is therefore very regrettable that long-term decisions concerning globalization are based on analyses that do not take them into account. In any case, we do not see how these phenomena would stabilize markets and reduce economic risk levels.

On the whole, it appears that the ID^3 model underestimates rather than over-estimates the consequences of instability phenomena that it helps to highlight. These phenomena are dramatic because they greatly damage the benefits expected from agricultural liberalization. Worse still, they change benefits to losses. Can we do better? This is what we are now going to examine.

---

[1] Newbery and Stiglitz (1981) advance the same reason in their famous book on the stabilization of staple food commodity prices.

CHAPTER VII

Could We Do better?

Checking price fluctuations and risks in a liberalized market

It is therefore unlikely that globalization benefits will live up to expectations. On the basis of standard, low quality models, by making use of comparative advantages, globalization generates significant resources, causes little frustration and improves the situation of all. But here, from a little more realistic model, it can be seen that market failures will squander such benefits and even completely cancel them out. Basically, this is due to two reasons:

The first is that, as a result of agricultural market instability, opening borders worsens price fluctuation. Price fluctuations are not at all "unpredictable", or triggered by phenomena outside the system such as vagaries of the weather for which insurance could be established at global level. They are "endogenous", an offshoot of the very functioning of the system, meaning that it will be difficult to get rid of them without reforming the system itself.

Hence, in particular, no insurance system can reduce the impact of such fluctuations: an insurance system covering the risks triggered by the functioning of the market could modify the functioning thereof. Thus, the magnitude of the insured risk will depend upon the existence of insurance coverage. Such is exactly the type of situation insurers resent, since it is a one-way ticket to bankruptcy. Now, contrary to a popular belief, insurers, as a consequence of the law of large numbers[1], never take any risk for themselves. Therefore, being exposed to bankruptcy from an endogenous risk generation process is certainly not their cup of tea.

The second reason for which liberalization generates unpleasant effects is that fluctuations have an adverse effect due to producers risk aversion. Indeed, producers resent risk. In face of a risk, they decrease production. That is why market liberalization, which would normally generate effects similar to those of technical progress by allowing for the exploitation of comparative advantages, behaves here just like a negative technical progress which could render production more difficult.

Denying agricultural exception, an unwise option with numerous consequences

This is not a generalized phenomenon in economics. It is specific to agricultural (and similar) commodities. To be more precise, it is specific to commodities characterized by low demand price elasticity, a peculiarity which, as a matter of fact, leads to the price instability we just referred to. Agricultural commodities are in this situation (once more, not all of them), and others as well! However, the others, medical services especially, are, in general, not easy to transport and largely insensitive to fluctuations in international

---

[1] Indeed, given a portfolio of contracts, the law of large numbers allows great precision in forecasting the amount of payments to be associated therewith, on condition that the contracts are numerous, that none of them is more important than the other and that the risks are independent of each other. Yet, as a matter of fact, with the type of price risk referred to here, even with numerous "small" contracts, the risks may, after all, not be independent... the law of large numbers does not apply.

markets[1], such that the issue arises only in relation to agricultural products.

As a result, and up to a relatively recent past, agricultural commodities were not taken into account in trade negotiations. From this viewpoint, there was great consistency between GATT (General Agreement on Tariffs and Trade) activities in liberalizing almost everything except for agricultural products and analyses of the situation by economists in the 50s and 60s. Setting aside this analysis led to denying the existence of an "agricultural exception" (Bouët, 2001) and resuming negotiations on that basis. For want of something else to liberalize, negotiation would definitely have lost its very *raison d'être* had agriculture once more not been taken into account. If this unwise stance were to be maintained, it would lead to high levels of price fluctuation around an average significantly higher than what obtains currently. There would be bankruptcy, ruin, farms would be abandoned and the shortage will force consumers to use a higher amount of their income on feeding. It is a foolish policy.

This does not mean that existing agricultural policies are satisfactory or cannot be improved. In agriculture, there are doubtlessly undue location advantages, awkward settlements and many other practices that need reform. The share of agriculture in modern State budgets cannot be considered as satisfactory. However, what has just been demonstrated is that the *laisser-faire* and recourse to the free market is surely not the right solution to remedy the inconsistent nature of agricultural prices.

The issue therefore is to know if it is possible to do better. First, could we not think of "liberal" solutions

---

[1] Also, many of these "basic commodities" are not at all governed by "market laws". Indeed, medical services, for instance, are far from being "governed by market laws" in civilized countries. The market nonetheless plays a non-negligible role therein, "on the sidelines", for example through the "freedom of establishment" of practitioners where they so desire.

which do not encourage the perverse game of price fluctuations? Could we not, to be more precise, envisage financial solutions like "derivatives" through which "income insurance" can be obtained?

As een above, the ID^3 model is yet to provide an answer to this question. However, we also realized that most of the results had already been provided by the risk cobweb template presented in Chapter II. It may thus be useful to make use of this "model of model" to study the likely consequences of futures markets and other instruments designed for coping with prices instability.

The Futures Market Option

For long, futures markets - the prototype of derivatives markets we just referred to – had been perceived as efficient means of avoiding the inconveniences of market instability. But what does this mean?

Futures are special markets wherein trade is not on physical quantities as in ordinary or spot markets, but on promises. There is a vast gamut of such promises. The simplest of them - indeed the only one that deserves the appellation of futures market - is the one that consists in the buyer promising to purchase a given quantity at a given date at a pre-determined price and the seller promising to sell the said quantity at the same price on the same date.

"Options" may also be sold. Where at such date (or during a given period,) the price drops below a price p_o, then I will promise my partner in the transaction to buy (or to sell) at the set price if they so desire (if they "exercise their option"). Otherwise, I am free of any commitment. An infinite number of other formulae are possible, all of which are, however, based on "contingencies". That is why it will be easier here to think along the lines of the simplest contract, referred to as the forward contract. What is going to be said in this regard is applicable with some variants to almost all contracts of this nature.

For ages, traders have practiced forward sale and purchase on the basis of mutual trust. Trust is important because, obviously, there is always a "loser", i.e. the person who promised to buy at 50 when current price is at 10, or to sell at 10 when current price is at 50. Naturally, it will always be *a priori* in the interest of "losers" not to respect their commitments. But this is not necessarily the case. In *intuitu personae* trade, losers have good reasons to honor their commitment. They have a high sense of honor; they know their partner well and are convinced that they can become the winners in subsequent transactions. The context could be different in a large anonymous market.

The innovation brought by the introduction of financial capitalism is the organization of a market around these forward securities, a market "policed" through various technical means to secure the successful outcome of transactions even in the absence of trust among participants. Thanks to this situation, traders who buy or sell in the futures market may seek their clients or suppliers within a wider circle than that of their personal relations. It is hoped that the price would be "more attractive" on this larger market. This is done without losing the major interest in selling forward within a smaller group: the possibility of using the method to hedge against price fluctuations. During the cropping season, a farmer may sell his harvest forward and hence be assured of the price he will reap therefrom.

The price in question is the market price, i.e. it is determined in a neutral manner without preference for any particular operator. It is imposed on each of them as an exogenous force. It is this "objectivity" that is valued in the price management mechanism which depends on no particular lobby and cannot be suspected of having been influenced by any behind-the-scenes manipulation. But what does this objectivity truly reflect?

Operators on the futures market can be classified into two categories: those interested in having high prices (farmers) and those interested in keeping them low (e.g. millers). Nevertheless, it is in the interest of both categories to have a stable price. Hence, to many, the price on the futures market appears to be a reflection of a mutually beneficial equilibrium value, since the risk factor disappears and is replaced by a "blue chip" which is advantageous to both parties. If reality was such, prices on the futures market would be stable and close to marginal production costs.

Unfortunately, this is not the case. Theory shows, and this has been confirmed by experience, that prices on futures markets are just as unstable as those observed on spot markets. Under no circumstances can futures markets "stabilize the prices". Rather, they allow to "take an insurance" against fluctuations. But there is a cost for such insurance. This is conditioned by risk aversion.

If I am a farmer, by operating on a futures market, at sowing time, I know exactly at what price I will sell the harvest that I believe I will have. As such, I can adjust my efforts taking this into account. Since the price will be certain, it presents none of the disadvantages of unstable pricing applicable if there were no futures market. Therefore, I need not seeking protection through a risk premium to reduce production with regard to what is technically possible. However, for this to happen, I need to find a correspondent ready to bear the risk I am not willing to take. (I could even do this anonymously through the market). It is unrealistic to think that the mythical correspondent will not require a reward. It is possible to meet people who are involved in gambling (casino or horse racing) in conditions where their profit expectation is negative. They like to take risks when small quantities or amounts are involved. It is rather unlikely that this "risk supply" will be capable of covering all the risks of the agro-food market of the planet[1].

Just as untenable is the idea that, by allowing demand and supply to project into the future, futures markets would, in theory, balance out the future marginal cost with the future price (Mocilnikar, 1998). Indeed, this argument is just as difficult to admit as the idea of free markets balancing long run marginal cost with current demand. This presupposes that the agents are at least aware of the future demand curve, which is not easier than for the current demand curve. It also presupposes that there is no speculator attempting to take advantage of errors made by other operators. If such considerations are included in the model, the latter's results will display the properties of the standard cobweb, with, in particular, a cost of the forward sale for the producer (See Box 12).

---

[1] The arguments just presented were made by Keynes. In technical jargon, they are known as the "normal backwardation" theory, since backwardation is the difference between the expected spot price for a future date t, and the forward price today for a transaction that has to be carried out at the same date t. It is "normal" in this case that backwardation be positive.

The normal backwardation theory has been subjected to empirical tests which were rarely conclusive. The research concluded that the theory was wrong. Another interpretation of the result is that, in reality, the futures market has never served as large-scale price insurance (see Williams, 2001).

Also, why would the normal backwardation theory be wrong? Because, according to some analysts, there are some operators who try to diversify their risk portfolio and take risks in other sectors uncorrelated to those found in their main portfolio. If, indeed, it were possible to ensure coverage by causing risk to disappear through a phenomenon close to that at the origin of the law of large numbers, then the argument will be faultless. However, as shown above – and as confirmed by all our simulation results – there is very little hope for the law of large numbers to be applicable in this case due to the fact that market parameters depend precisely on risk taken by the business.

> **BOX 12   THE ALGEBRA OF FUTURES MARKETS**
>
> Let us imagine a farmer who plants wheat and a miller who wants raw material for his mill. If both of them were "risk neutral" (indifferent to price dispersion), they would made their plans on the basis of the expected price, equating marginal costs and prices "on average". In that context, the futures market would be useless. Yet, if the market does exist, it is because "there is something else". This "something else" is first the risk aversion of both the farmer and the miller, and, second, the existence of speculators. The risk aversion induces both farmers and millers to secure the prices they receive or pay. The speculators try to make profit from that circumstance. They offer warranted prices to both farmers and millers, promising at time 0 to sell or pay wheat $p_{0\tau}$ at date τ. But for them to make a profit, $p_{0\tau}$ must be less than $\bar{p}_{0\tau}$, the expected price of wheat at date t given the information available at time t = 0. The speculator average profit will then be $\bar{p}_{0\tau} - p_{0\tau}$. It is the risk premium known in actuarial jargon as normal backwardation. It represents the "insurance cost" – technically speaking.
>
> Is this risk premium high? There are few studies on the matter. Some of such studies posit that it is very low, while others find it significant. The clear difficulty researchers face over the issue is that of determining $\bar{p}_{0\tau}$, the expectation at date 0, in view of available information, of the price that can be imagined for date t. It is common knowledge that this expectation is different from the spot price p^*_t as could be observed *a posteriori*, since new information would have reached the market between date 0 and date t. What could be done to determine it when, by definition, it can under no circumstance be observed? That is why all attempts at evaluating the "futures market cost" have never really been accepted by all economists. Trying to do so correctly would be tantamount to trying to understand the meaning of chance, an age-old philosophical problem that is far from being resolved!
>
> What is certain is that the farther the due date, the faster the cost increase. According to the most optimistic theory in this regard (the "random walk hypothesis"), variance increases proportionally to time. It is believed that such are the main reasons why futures markets do not exist for the long term, since the cost of risk becomes simply prohibitive. Indeed, it is rare for farmers to forward sell effectively before planting. The most common markets are "six months off", while farming term is one year, or nine months at best. This significantly prevents farmers from using futures markets.

Fluctuations reduce production in the same manner as reverse technical progress would

In fact, it can be easily demonstrated that for farmers, by eliminating premium (which need not come into play here), the expected forward sale cost is barely equal to the expected profit of speculators, which is also, on average, equal to the risk premium of speculators. There is no reason to expect this profit to be negligible in the long term, unless one were to admit that speculators are involved for the sake of it, which is simply unlikely. However, what is important is that from that perspective, the use of the futures market is described as "neutral on average" for the community, since the losses of farmers are just equal to the gains of speculators. As a matter of fact, this is profitable in the sense that the farmer (poor) gains more security and the speculator (rich) makes profit, like in the story of the beggar by Daniel Bernoulli (Box 2, Chapter I).

Yet there is more, and this leads us to the second point to consider: our concern is not really to evaluate the risk premium thus transferred from "producers" to "speculators" through the futures markets. The real issue is to find out if these instruments could reduce fluctuations and remedy the fact that fluctuations, as a result of the risk premiums subscribed by producers, could diminish production in the same manner as reverse technical progress would. From that perspective, we see the conclusions change completely once more, depending on whether we are considering the case of an "exogenous" or "endogenous" market fluctuation.

In the case of exogenous fluctuations, and with a variance independent of the volume produced, it is obvious that improved security on the part of the producer will lead to increase production, and hence, in the long term, induce a drop in the average market price, to the benefit of the consumer. Under certain circumstances, the overall variance could even decrease. However, if as seen above, the variance, more than the average price level, is used in transmitting information between producers and consumers, then it is very likely that the existence of a futures market will, instead of reducing fluctuations, rather create some where there were none (Boussard, 1991)[1].

---

[1] It should be noted that the only thing that is certain in this hypothesis is that we know nothing: within the framework of the parameters of

Stocking and Destocking

Another standard way of overcoming market inability to reach equilibrium is to stock or destock.

In principle, State intervention is not needed. Just anyone can understand that it is possible to make profit by buying when prices are low and selling when they are high. In so doing, one encourages the market to return to equilibrium. As such, there should not be fluctuations in the market for a product which can be stocked. Since out of experience we know this is not true, there has to be a shortcoming in this line of reasoning. Indeed, the shortcoming is not difficult to spot: the reason is that speculators make mistakes, sell at a low price and buy when prices are high, thereby increasing shortage and rousing the people's anger against "monopolists". As a matter of fact, speculators are never certain of the price at which they will resell, and consequently, they take a risk. As a result, they find themselves in exactly the same situation as the agent on a futures market, and for the same reasons: to such extent that both practices, stocking and destocking, are often compared with futures market operations. Of course, when situations are similar, problems are identical. All that has just been said about futures markets also holds for stocking and destocking.

Could State intervention correct the well-known difficulty markets have in "thinking long term"? As a matter of fact, things are a little more complex than that,

---

the model, the "chaotic" or "non chaotic areas" do not form sets of adjoining points, with simple geometrical forms such as bands, triangles or ellipses. These are "fractal" sets with "voids" here and there, such that it is impossible to forecast in a simple manner what parameter values could (or not) lead to a convergent or chaotic regime. Under such conditions, it is quite possible that the introduction of a futures market could contribute to turning a "convergent" market situation into a chaotic one, in view of the value of the slopes of supply and demand curves, and of risk aversion.

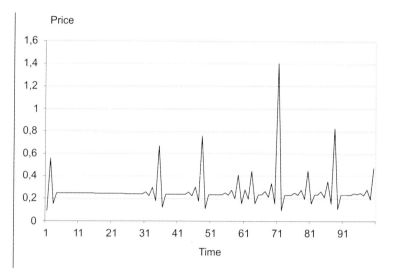

Figure 29 A chaotic cobweb model with a stock and destocking policy (source: Matlab Accurs function [available on request from authors] with the following parameters: Pmin = 5.1 [intervention purchase price]; Pmax = 8 [intervention selling price]; Al = –1.5 [demand curve slope]; Bet = 1 [ordinate producing the demand curve]; Af = 0.4 [fixed factor share in the production function]; Gam = 0.1 [annual depreciation rate of the fixed factor]; Vc = 5 [quantity of initial variable factor]; kc = 20 [quantity of initial capital]; Ss = 0.5 [producer's propensity to save]; r = 0.7 [capital resale price], Pk = 10 [capital price]; Lc = O [current sales of capital]; Pa = 20 [expected constant price].

as the issue at hand[1] is one of limited stocks. Hence, figure 29 presents the results of the same function, exactly, as that used above to construct figure 5 (Chapter II), but including a "state" stocking and destocking system, through which stocks are build up when prices drop below a minimum and destocking is done when they reach a maximum. The "free" market operates within the limits thus defined. There is, however, a maximum stocking capacity. Where the capacity is exceeded, one returns to the "free market". The same applies if the stock is nil. In addition, the producer has the possibility of borrowing, if not to produce, at least to survive, where he had incurred losses.

---

[1] This problem was well understood by Newberry and Stiglitz (1981), yet in a context of "Gaussian random walk hypothesis," rather simplistic compared to models considered here.

This result shows the adverse effects of stocking: nothing happens for a long time, and then, someday, the stocked quantity becomes nil and the result is disaster[1]: prices skyrocket to unexpected highs, ruining consumers and enriching producers[2]. The system nevertheless turns out to be rather efficient and comfortable for the consumer. It however presupposes heavy and costly State intervention.

A fundamental point in interpreting the above result is that the system described above still leaves a large room to price fluctuations – and their adverse effects. It is only because the range of intervention is large that the system does not "derail" towards a permanent saturation of stocking capacity. If the range is too narrow, with an "homogenous and of degree one"[3], production function,

---

[1] Laroque and Deaton (1992) demonstrated the same phenomenon in a completely different context, with exogenous disturbances and speculators with "rational" (though imperfect) expectations. What is shown here is that there is no need to bring in an exogenous disturbance to obtain this result.

[2] It should however be noted that in many third world situation, producers and consumers are the same, at least in part. Thus, these phenomena cannot be approached solely from the income redistribution perspective.

[3] A production function which is "homogenous and of degree one", or better still of "constant return", is such that if input quantities are multiplied by any number, the corresponding output quantity should also be multiplied by the same number: as such, if it is possible to produce 10 tons of wheat on 1 hectare of land, with 20 hours of work, 350 kg of fertilizers and 3 hours of a combine-harvester, it is therefore possible to produce 20 tons on 2 hectares, with 40 hours of work, 700 kg of fertilizers and six hours of a combine-harvester. This shows clearly that in the case of competition, the average production cost and the marginal cost are exactly equal to the price, and that the marginal cost is constant. Under such conditions, without fixed factors, if the price system is such that the cost is less than the price, production will be nil. If it is higher, production shall be infinite. We thus swing directly from one situation to the other. Production functions in agriculture are usually homogenous and of degree one. That also accounts for agricultural price instability. In practice, the existence of fixed factors reduces the brutality of the phenomenon. However, there is no fixed factor in the long term…

as soon as the minimal guaranteed price is slightly above production costs, there is no limit to the increase of the quantity produced. That is indeed the reason why guaranteed prices without limitation of quantity, both in the United States and in Europe, led to the excesses we all know. Things could not have been otherwise.[1] What, in this case, gives flexibility to the system, is the existence of a rather high risk premium, fed by non negligible fluctuations. There is somehow a phenomenon of "optimal volatility" big enough to curb production and small enough not to lead to disaster, or to lead to disasters happening at long intervals.

Such intervention schemes have proven that they were possible in the past. For example, Thailand's rapid emergence as a rice exporter was a result of such policies, although stocking was replaced by the opening or closing of frontiers through a complex system of export or import licenses granted to importers and exporters[2]. In the same vein, before 1992, Argentina regulated the domestic price of grain through an exports tax which varied depending on the economic situation.

However, in practice, these systems are difficult to set up. Indeed, in the examples just mentioned, it has been noticed that the basic stocking role was played by the external world, not silos in the countries themselves. The latter would surely have been difficult to be constructed and maintained. From a purely egoistic viewpoint, taking advantage of the external world is not very fair. If it is possible for a "small country" to jolly well play the

---

[1] This was also the major criticism Colin Clark made against the common agricultural policy. He was doubtlessly the most foresighted agricultural economist of his time. In this regard, it should be noted that Colin Clark had nothing against the idea of dissociating agriculture from the market. What seemed simply impossible to him was supporting prices without limitation to quantity (see Peter, 2001).
[2] See, for instance, Kajisa and Akiyama, (2003)

game, from a Kantian perspective of international relations, it is obvious that this is not possible. The world should be considered as a closed system and therefore the stocking capacity should be envisaged in the world as a whole. That would be a major enterprise.[1]

Production Quotas

The last stabilization techniques formula to be studied here is the production quota. Through the introduction of a production quota, a given individual producer is guaranteed a given price (generally high) for a given quantity. If the producer produces less, he will receive the said price for the produced quantity. Where more is produced, depending on the case, the excess production will be destroyed or bought at nil price (which is what is currently going on in the European Union regarding milk) or better still it could be sold but at a international market price and for exportation only (which is what was going on with EC sugar until recently). The guaranteed price itself can take various forms.

The domestic market could be operated at that price. No quantity can be bought at a different price on the domestic market. This presupposes a system of variable taxes at the borders so that imports can be sold locally at the administered price. The producer support policy is thus financed by the consumer and does not cost a dime to national budget (which, on the contrary, swells up with customs revenue if any).

It is also possible to practice a system of deficiency payments, whereby the domestic price is the world price. Producers sell at that price (which is unavoidably

---

[1]See criticisms made by Newbery and Stiglitz (1981) against solutions of this nature. Yet, Newbery and Stiglitz considered a situation that was more favorable to stocking than the one described here, since they pay no attention to endogenous risk causes, assuming all shocks were exogenous, caused by climate-related or similar phenomena.

fluctuating and often low). However, upon presentation of invoices and within the limits of the quotas allowed, the Treasury refunds the difference between the obtained and the guaranteed price. This system is financed by the taxpayer, which is certainly preferable in an income redistribution policy perspective.

There is therefore an array of highly variable modalities for setting up a production quota system. Yet, nothing has been said about the variety of conceivable solutions for the allocation of property rights on the quotas. These rights may be granted to the land owner and be inseparable from landed property: a given piece of land is granted the right to produce such number of kilograms of a given commodity at guaranteed price. The quota may also be attributed to the producer, e.g. the farmer, and not the land owner. That is what obtains for milk quotas in Europe. The rights may be personal and therefore can be returned to the government when the holder ceases to farm[1]. They may be sold to a third party on a market, as is the case with milk quotas in Quebec.

Whatever the variety of such modalities, the idea of imposing production quotas is rejected *a priori* by most professional economists. Indeed, in a static analytical framework, that is, if one were to leave aside all consideration of motion and time, such a method looks senseless, imposing the burden of attributing producers an economic rent (which is fully comparable to a monopoly rent) either on consumers or taxpayers. At best, in response to the preceding remarks, an orthodox economist could accept the idea that since the market is not able to reach its equilibrium price, the State should

---

[1] Such is the case, at least theoretically, with milk quotas in France. In practice, the government does not always exercise its right of repossession and, *de facto*, the quotas belong to the farmer who sells them more or less legally.

play the role of Walrasian[1] "auctioneer" and determine the equilibrium price[2]. As a matter of fact, where the economic rent associated with the quota is nil (because the guaranteed price is exactly equal to off quota rent production cost), there will be no more disadvantages to the quota. However, by the same stroke, it will no longer serve any purpose.

This is true within a static context, in the absence of real uncertainty on the equilibrium price. In a dynamic context where the equilibrium price is unspecified (difficult to know in any case), this is no longer the case. Figure 30 shows the results of a model similar to the one in figure 5 (Chapter II). This was obtained with the equations of Box 6 (Chapter II), which were modified to take into account the existence of a production quota system as a result of which, in any event, the quantity under quota is always produced[3]. The dark curve represents the quota-free market, the grey curve a system of guaranteed prices for production under quota - a price 10% above the initial equilibrium price is guaranteed for producers for quantities below 10% of the initial

---

[1] Leon Walras (1834-1910) originated the idea of representing the economic world through a system of equations. He could be considered as the forerunner of the models referred to in the preceding chapters. He foresaw the difficulty of representing expectations, which was discussed in Chapters II and IV. He thought the problem could be overcome (as was done in the standard model) by imagining the intervention of the "auctioneer" who was expected to have organized an auction sale of all the goods and factors of production, with transactions intervening effectively only after a balance has been reached. It was an ingenious but hardly realistic idea.

[2] This is what was attempted in the case of agricultural policies in developed countries after World War II, with the policy of "guaranteed prices". The difficulty in this policy arose in fact from the inability of the State to know the exact position of the equilibrium point.

[3] The major difference stems from the fact that no hypothesis was made of a constant expectation for the current price. What was done was a "naïve" expectation of such price equal to the equilibrium price of the previous period: $\hat{p}_t = p_{t-1}$. The expected variance is therefore given by $\sigma_t^2 (P_{t-1} - P_{t-2})^2$.

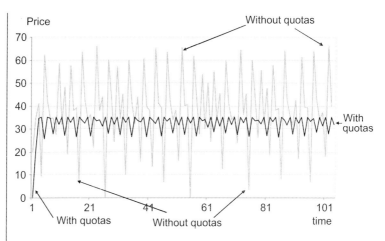

Figure 30 Effects of the introduction of production quota in a chaotic cobweb model.

In the without quotas situation, the average consumer surplus is 146, and the average producer surplus, -166. With quota, the cost for the government is 44, the consumer surplus is 109, and the producer surplus is 45.

Sources: Matlab CR3 file, available on request with the following parameters: "naïve" expectations; prices guaranteed by deficiency payment (for quantities under quota), supply curve slope, 3; ordinate of the start: 30, demand curve slope, –0.4; ordinate of the start, 50; initial quantity, 11; risk aversion, 0.001, expected equilibrium price, 20; quota in % of equilibrium, 90; guaranteed price in % of equilibrium price, 110.

equilibrium quantity.

Results speak for themselves: while the non regulated market leads to large-scale chaos, the markets with quotas also lead to chaotic fluctuations though of a much smaller magnitude. The cost for the government is relatively lower than the social benefit. In the case under study, the cost for the government is at least theoretically recoverable through a tax on producers[1]. The EEC sugar

---

[1] With the data in figure 30, a tax of 90 on producer income (leaving them with a profit increase of 70), allows to refund government for his expenditure of 45 and to grant consumers tax rebates of 45, hence compensating, and beyond, their 30 excess loss. It should be noted that producers' risk aversion here is low enough for us to find ourselves in the situation described by Oi (1961), in which "consumers benefit from" instability. In reality, it is clear that producers will not continue to produce for a hundred years with this kind of losses!

market, until recently, was organized in that manner[1]. The fact that nobody in Europe complained after the high price of sugar shows that it was operating perfectly and to the satisfaction of all. Had such not been the case, due to the volatility of the sugar prices on free markets, this commodity would have been on the headlines.

Naturally, the implementation of a system of this nature cannot be done in a haphazard manner. The quotas would create rigidities in production and generate economic rents which should not exceed certain limits. However, this type of disadvantage can be checked by allowing producing rights to be exchanged on the market, as well as by determining the quantities and guaranteed prices at reasonable levels.

As seen above, if price and quota are set at equilibrium levels, all the disadvantages of quotas, (as well as their advantages) vanish. Obviously, the difficulty in quota administration and "rational expectation" is to determine the level of equilibrium. Managing such uncertainty through a quota system makes it possible to overcome the difficulty: the quota could be set at a level known to be below the equilibrium and the market allowed to deal with marginal quantities.

A flexible quota system enshrined in a market economy is, after all, an old idea of American rural economists[2]. The idea was never applied. Presumably, in the United States like elsewhere, this is due to the fact that economists are hardly taken seriously by governments, except during very severe crisis, or very

---

[1] However, the European Union sugar system does not include deficiency payments: the stabilization cost is totally paid by the consumer in the form of high domestic prices. Incidentally, the budget cost of the system is nil.
[2] See for example Hazell and Scandizzo (1977)

belatedly. There is however another reason for this: neither economists nor government have ever seen the profound similarity between the quota and the futures market systems - hence the almost complete compatibility of quotas with a perfectly orthodox market economy.

Are Futures Markets and Quotas Different in their Principles?

Indeed, let us develop an idea which, although *a priori*, shocking for someone raised in the religion of the market, nonetheless deserves to be examined: there is a close similarity between the principle of production quotas and futures markets, because the system of flexible quotas described above could be interpreted as a forward sale to the State of the quantity under quota at harvest time, the State eventually bearing the risk associated with the resale at spot market price.

If there were operators on the market ready to buy at the guaranteed price a futures for the same quantity as the one put under quota, everything would be done as in a system of effective quotas. Of course, such operators do not exist, first because it is difficult to mobilize the huge financial resources required, second and more importantly, because private operators with no inclination to get involved in speculation with negative profit expectation will be almost sure to lose "on average" the amount of the economic rent associated with quotas.

Yet, in the absence of speculators, it might be in the interest of consumers to team up to pay the "risk premium" in order to secure supply. This premium could end up being far below the losses resulting from market uncertainties. Nonetheless, they would have to organize themselves in this regard, which is not very obvious. It would be easier for the State to assume this organizing role.

There is another reason for this role to be entrusted to the State. Theoretically at least, the ideal price for quantities under quotas would be the price which will simply turn the economic rent associated to the quota to nil. Then, the entire production under quota would be sold on the market at the guaranteed price. The latter would be exactly equal to the marginal production cost, while the rent would be minimized as if by an ideal market. The reason why we mentioned above that the economic rent associated to the quota ought to be always positive is that, indeed, no one knows exactly what quantity should be put under quota, or, even so, what its marginal cost is. For security reasons, it is accepted that the guaranteed price should be a little higher and, as a result, the corresponding rent should be positive. It is however clear, theoretically at least, that the rent could be considered as nil in the first approximation.

Under this nil economic rent hypothesis, a risk neutral speculator (that is, maximizing mean expected profit without consideration of volatility) could consider operating on the market. However, generally, speculators are not risk neutral, as we have seen above. They cannot be contented with an expected zero gain. They need a positive gain expectation to counterbalance the risk they take. Consequently, to propose this transaction on a futures market, the producer price should be clearly lower and the consumer price higher than the "equilibrium" marginal cost referred to above.

On the other hand, if there is any such thing as a risk-neutral operator, it should be the State, the only economic entity capable of withstanding very high risks without any significant disadvantage[1]. In addition, it is the general interest of all citizens that the State bore the costs at the actuarial value of the major non insurable risks

---

[1] This point is rigorously demonstratedin many theoretical works on the role of the State in the economy, especially by Arrow and Lindt (1970).

without asking for premiums as private operators would. That is a standard theorem in public finance. That is indeed why, in the long run, a flexible quota system such as the one just described is very close to the "ideal futures market".

It should be noted that in such a hypothesis, speculators could still play on the prices of the "producing rights" which would have to be traded on a parallel market (authorized or not), as seen above. In fact, producing rights prices would fluctuate depending on government policies (depending on whether the government set the guaranteed prices at a high level or, on the contrary, causes them to drop for consumers to benefit from technical progress), population trends (they would rise if many young farmers desired to settle) and other circumstances. They would behave in the same manner as the price of land, which is another form of quantitative limitation to production.

The advantage of a futures market on production rights, which will not deprive speculators of their potential benefit, is that contrary to the case of the futures market bearing directly on the quantity of goods traded, this one is completely neutral with regard to production, whose volume is guaranteed irrespective of what happens. The only likely disadvantage of such a configuration is that the average price of the producing right would then be lower than the capitalization of its expected gains, due to the existence of a risk premium for speculators.

Holders of "historical" producing rights could complain about it, whereas the newcomers who settle, and who have to buy rights to produce, would be happy. However, the interest of the consumer would be protected no matter the situation, whoever the final beneficiary of the economic rent or risk premium turns out to be. It can clearly be seen that what matters here is not the producer, but the consumer.

To Conclude

These remarks are somewhat out of the main concern of this work. However, we could not have examined the disadvantages of liberalization without proposing a few possible alternative solutions. We are well aware that the proposals are imperfect, sketchy and incomplete. Moreover, the economist should not play the role of a public administrator by developing institutional solutions which are not in his sphere of competence. He may however suggest avenues for research, which is what was done here.

Obviously, "all market" solutions cannot work in the case of agricultural products. This does not mean that the market is totally useless. Though it is far from possessing the more or less magical powers[1] that the "free trade" propaganda grants it, it has a major role to play, "on the sidelines", to eliminate the problems relating to the inflexibility of "bureaucratic" solutions and centralized planning. It lubricates the mechanism, which is necessary for the system to operate properly. Such is the lesson to learn from this chapter, hoping it will be taken into account.

---

[1] A program recently designed by the American federal anti-terrorism agency clearly demonstrates the magical attributes given to the market. Its objective being (amongst other things) to identify areas where terrorists are likely to strike, one of the solutions proposed consisted in setting up a futures market where bets on the next target would be traded. Based on the fact that operators would receive payments depending on the quality of forecasts obtained, the promoters of this lofty idea believed that the "market would bring out useful and relevant information". Obviously, one simply has to look at thinks from the point of view of a terrorist in search of a target to realize that the latter would avoid attacking the best guarded targets : here would be the source of chaotic motion in terrorism targeting!

CHAPTER

Recent Developments

The bulk of the present book has been written in 2004, based on studies made in 2001-2003. Some time elapsed since. Are recent developments of the world agricultural markets confirmed or infirmed the above conclusions ?

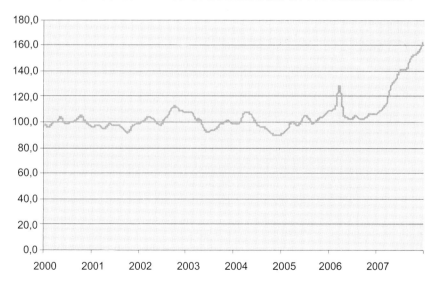

Figure 31 Index of actual imported food commodity price in France, 2000-2008.
Source: AGRESTE, IPPAP, matières alimentaires importées en euros

The most obviously surprising recent development in the world agricultural economy is the upsurge of prices in 2006-2007. As an illustration, figure VIII-1 shows the price index of *all* imported agricultural commodities in France since 2000 – a reasonable proxy for the whole world traded commodity price index. From a 100 reference value in 2000, it now (early 2008) reaches about 180. If one take into account that this index is computed in current Euros, and since the rate of exchange between the Euro and the US $ has recently been increased by about 50%, the same index in dollars would have largely cross over the 200 % limit.

Now, if most "standard models" have predicted a slight increase in agricultural prices, as a consequence of liberalisation, none of them ever envisaged international commodity prices being multiplied by 2. Quite the contrary, many analysts, until 2005, were complaining about the "iron law" along which agricultural prices were declining, hurting poor farmers in developing countries, and depriving them of any chance of breaking the cover of the poverty trap. Even when a slight price increase was envisaged by a few analysts, (see figure 32, derived from the FAPRI model of the Iowa Sate University, baseline 2002), it was a gentle growth – nothing common with what can be observed on figure 31 – 33.

To date, broadly speaking, the ID^3 model results are fairly well supported, although in the details, discrepancies between model and reality do exist, and although, in early 2008, ID^3 predicts a turning point which does not materialize yet. Of course, in the near future, such discrepancies could exist also, and even be larger. Let us examine a few results.

The model has been used in 2005 to investigate the ex post consequences of the changes in the European Common agricultural policy (CAP) in 2003, and the ex ante consequences of a further contemplated change envisaged for the near future. Three scenarios were built up accordingly :

a) A baseline scenario, assuming the continuation of the CAP as it was in 2000, involving, in particular, "intervention" prices, that is, a floor price at which the Government buy any commodity offered on the market, as well as a ceiling price, at which Government takes measures to avoid a further increase.
b) A "reform" scenario, corresponding to the reform effectively enacted and put in effect in 2003. It consisted for the most important part in suppressing the intervention price, relying on decoupled subsidies to help farmers.
c) An "opening" scenario, assuming remaining duties and price management institutions will be suppressed – in particular, all export subsidies.

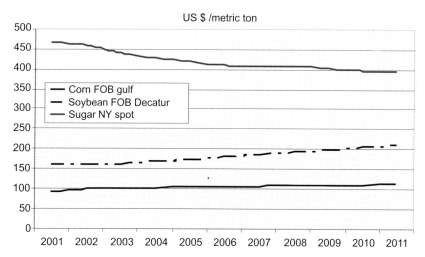

Figure 32 FAPRI baseline 2002 price forecast
Source: FAPRI, Iowa State University, baseline 2002

The corresponding results for the price of US wheat in these three situations are displayed on figure VIII-4 :

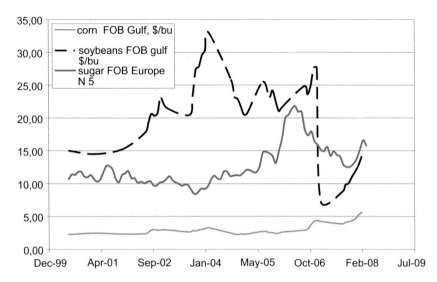

Figure 33 Actual development of the FAPRI commodity prices
Sources: For wheat and soybeam, USDA agricultural outlook archives, table 19, prices of principal US trade products; For sugar : London stock exchange, table 2, World refined sugar prices monthly average

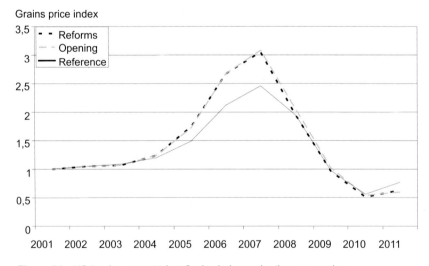

Figure 34 US "grains except wheat" price index under three scenarios
Source: ID3-C:\nov07-minag\LAG_GTAP01-7reg-bas99-janv08.gms; C:\nov07-minag\LAG_GTAP01-7reg-reformesjanv08-T12.gms, and C:\nov07 minag\LAG_GTAP01-7reg-ouverture-sjanv08-T12.gms

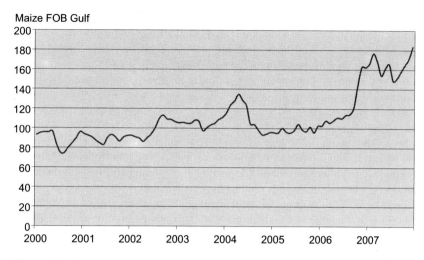

Figure 35 Actual price index of Maize in the US, 2000-2008
Sources: USDA US No.2, Yellow, U.S. Gulf (Friday), US$/ton

A sharp peak is forecasted for 2007, and a decline after, followed by a deep hollow, which makes the 2010 below the 2001 price. This is more or less what can be observed on the actual price index of US maize, as displayed on figure VIII- 5. While the actual timing might be slightly different - most analysts are now predicting a 2008 price significantly lower than 2007, not perceptible on figure VIII-5 data – and despite the failure in representing the temporary small high point in 2004, the model is quite similar to reality. Indeed, the model only slightly exaggerate the magnitude of the peak, which is 2 instead of 3. This is the more striking as when it was presented in 2005, the most common comment on this curve was simply "It is incredible !".

As a matter of fact (but this is anecdotic) it shows also the importance of the 2003 CAP reform, which actually changed the whole world food economy. The peak would have been less pronounced with the "reference" situation, in the absence of the CAP reform in 2003. By contrast, the further liberalisation projects are perfectly innocuous, at the point that the two "reform" and "opening" curves are almost indiscernible. This is because, before 2003, the CAP was practically without protection from foreign imports, while still maintaining a minimal price regulation. In this way, it served as a regulator of the whole system, buying excess supply at almost fixed price, and destocking when prices had a tendency to increase[1]. The model here confirms an hypothesis which have been expressed by various writers: the positive role, at global level, of regulation policies put in operation by a few big producers, as the EC or the US (at the expense of their taxpayers, but for the benefit of themselves, and of the whole world community also).

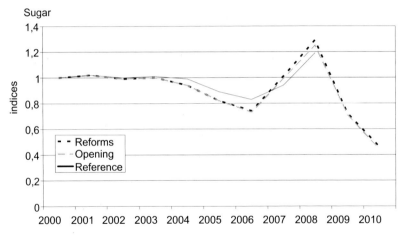

Figure 36 Sugar developing countries price index under three scenarios
Sources: LAG_GTAP01-7reg-bas99-janv08.gms; LAG_GTAP01-7reg reformesjanv 08-T12.gms, LAG_GTAP01-7reg-ouverture-sjanv08-T12.gms

A similar conclusion can be derived from figure 6, displaying the sugar price index for the "less developed countries". In this case, ID3 (just as FAPRI, and other similar standard models) envisages a lowering of sugar price at the beginning of the period[2], followed by an

---

[1]Delorme (2007)

increase in 2006-2007. In effect, it is conformable with the observed sugar price tendency, as pictured on figure 33.

More importantly, the comparison of the actual and forecasted price curves shows also that the role of bio fuels does not seem to as important as most analyst are believing. If the demand for bio fuel had been determinant for explaining the observed scenarios, the price of sugar should have increased, following the increase of the flexuel equipped Brazilian motor cars. This is not what happens with the ID3 forecast – this is normal, since ID3 ignores bio fuels – but it is also the case with the actual series, which does not seem to be much affected by this change in the general economic environment. By contrast, the 2003 EC liberalisation policy seems to have been efficient in lowering the international sugar price, as shown by the difference between the "reference" and the "reforms" scenarios. It also increased the height of the peak after recovery.

Figure 37 US wheat price index in ID^3
Sources : LAG_GTAP01-7reg-bas99-janv08.gms; LAG_GTAP01-7reg-reformesjanv08-T12.gms, LAG_GTAP01-7reg-ouverture-sjanv08-T12.gms

In these cases, the ID^3 model performs relatively well, despite the fact that the high prices should now begin to decrease, which does not seem to materialize at writing

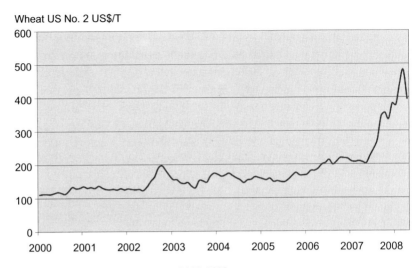

Figure 38 Actual US wheat price, 2000-2008
Sources: FAO commodity and trade division web site

time (April 2008). This is not to say that it performs well at all time and for all commodities. For instance, apparently, it does not provide a good forecast for wheat: In all scenarios, as shown on figure 37, it decreases from 2001 to 2007, instead of increasing (yet, it increases again after), as in reality (see figure 38). Nevertheless, the general shape of this curve is certainly more akin the actual development of wheat price than it is the case for FAPRI forecast.

Again, the contrast between the EC "reforms" and the baseline is striking. Once again, it shows the impact of the European policies on world markets.

But for lack of place (and the fear of boring the reader), many other similar graphs could have been displayed, in particular with meat prices. They broadly reflect the present situation, with the same accuracy (and lack of accuracy) according to contemplated commodity. Of course – and this quite general a remark – drought in Australia or elsewhere do not play any role in explaining ID^3 prices increases: it is thus completely wrong to attribute the latter to the "hand of God". Indeed, it must be attributed to the irresponsibility of decision makers who, advised by myopic economists, decided to suppress all the barriers which had been built during the forties and the fifties in order, precisely, to avoid such a situation.

The result of these policies is exactly what was predicted: increased price volatility, higher prices, and consumer's loss of utility – especially, in the case of the poorest. In fact, not all the poor suffer in the same way: the urban poor, who must buy their food, are the most affected. The rural poor, usually, do not need passing through the market, and are therefore relatively immunized against price vagrancies. Yet, increasingly, in many developing countries, the lack of land forces the emigration of many poor peasants toward the urban Eldorado. In this context, the present increase in food prices is larger tragedy everyday.

It is contended that the sudden rise in world agricultural prices will help the poor peasants, providing them the opportunity of increasing their incomes by selling their products. To a small extent, this is true. Some of them will benefit the windfall. Yet, they will not be numerous, and the windfall will be small. It would be large only if they could increase their productions largely beyond self-sufficiency. But this is precisely what most of them cannot do, for lack of capital. And since nobody (especially, no banker!) is persuaded enough that the present situation will last, nobody will lend them any money to be invested. In this context, the best they can get from the conjuncture is to avoid the direct consequences of food price rise.

Figure 39 illustrates the point: In 1994, the moneys of a few west African countries were attached to the French Franc by a fixed parity. Since the latter, at that time, was a "strong" money, they were recommended to devaluate, in order to get a price advantage on export. Overnight, in January 1994, the price of all foreign goods was multiplied by 2. According to common wisdom, such a

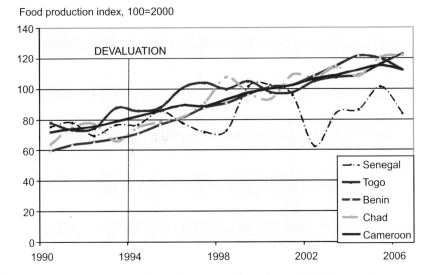

Figure 39 Actual food production index series for various West African countries

shock should be visible on macroeconomic data. But this is not true, for the reason indicated above : for lack of investing capacity, no producer could seize the opportunity to change plans and benefit from the new situation. Actually, whatever the country, no immediate or delayed effect of the 1994 100% change in foreign exchange parity on food production is perceptible, despite expectations from international organizations. Unfortunately, the consequences of the recent upsurge of food price will probably have the same effect on the agriculture of the poor, for the same reason.

It will benefit rich farmers (even the rich in poor countries), but only for a short time, since, if our forecasts are correct, it will not last for long. When prices will decrease,, then, the rich farmers will also suffer, without any profit for anybody.

Therefore, nothing must be changed in the conclusions of the preceding chapter, regarding the benefit of a global or national agricultural price regulation, and its feasibility.

Conclusion

> *Whether they are right or wrong, ideas rule the world*

In the conclusion to his celebrated *General Theory*, John Maynard Keynes maintains that, "right or wrong", ideas rule the world[1]. Undoubtedly, he was right: politicians' current concern to "liberalize" everything that appears to lend itself to liberalization obviously derives from lessons taught in faculties of economics and specialized research institutes. However, it is not certain that such theory is "right" always and everywhere – particularly in the case of agricultural commodities that are at the center of contemporary trade negotiations.

---

[1] "The ideas of economists and political philosophers both when they are right and when they are wrong, are more powerful than is commonly understood. Indeed, the world is ruled by little else. Practical men who believe themselves to be quite exempt of any intellectual influence, are usually the slaves of some defunct economist madmen in authority who hear voices in the air, are distilling their frenzies from some academic scribbler of a new year back. I am sure the power of vested interests is grossly exaggerated compared with the gradual encroachment of ideas. Not, indeed, immediately, but after a certain interval. For in the field of economics, and political philosophy, there are not many who are influenced by new theories after they are twenty-five or thirty years of age, so that ideas which civil servants and even agitators apply to current events are not likely to be the newest. But, soon or late, it is ideas, not vested interests, which are dangerous for good or evil". (Keynes, 1936)

It may well turn out that the theory according to which "since dirigisme has failed, liberalism is the future" is just as wrong as its contrary, "capitalism is dead, long live central planning", that communism, in its time, sought to impose[1]. Besides, the reasons for which such groundless ideologies may seem to triumph at a given time are related: they are based on simple ideas, often partially right, and easy to "sell" to uncritical decision makers.

But over and above slogans, the real economist has to analyze situations, understand the motivations of others and deduce therefrom how initially incompatible ideas end up engendering global situations that we witness. This is what we sought to do in this book basing on the microeconomic theory to "marginally" modify the standard macroeconomic model, so as to render it a little more realistic.

Admittedly, we did not complete this endeavor which is, by the way, gigantic. But the little that we covered throws some new light on the consequences of agricultural liberalization. The latter are very far from auguring the bright future promised by advocates of tariff dismantling. By adding a pinch of uncertainty to the mechanics of models generally used to evaluate the benefits of liberalization, we realize that the latter is much less beneficial than it was believed to be. Of course, it allows for better use of production factors and therefore, achieves efficiency gains. However, such gains are largely offset by the losses associated with price fluctuations arising from the functioning of the market itself. We can therefore affirm that agricultural liberalization will not necessarily generate a significant increase in production, induce development in poor countries, significantly improve income distribution in

---

[1] Today, one should reread with care Joseph Schumpeter's extraordinary book, *Capitalism, Socialism and Democracy* (Schumpeter, 1942), which not only denounced the emptiness of Marxism which was then triumphant in many American and English universities, but also proposed an "operating procedure" for capitalism with a human face.

the world and cause a decline in the prices of food products to the benefit of consumers, as is usually promised. It shall instead generate just the opposite effects.

This is not because liberalization is good or bad "in itself". Such an affirmation is absurd, in both senses. This is because, in the case of agricultural commodities, conditions are not conducive for the market to permit efficient exploitation of "comparative advantages". Consequently, a substitute for the market has to be found, or rather the market has to be managed by institutions that enable it to play its normal role namely, giving price its status of real messenger between consumers and producers.

In this regard, we have made some proposals probably quite cursory and incomplete. They are more research proposals than real solutions. One notes that though they are based on some kind of mistrust of market mechanisms, they are not incompatible with extensive participation of the market in establishing the necessary equilibriums.

Why not explore these avenues instead of limiting oneself to the search for an equilibrium which, though "natural", is nonetheless problematic, and though "efficient", is not necessarily "fair"?

As did the Marxists in the 60s, one may imagine that there is somewhere an international conspiracy dedicated to ruin mankind. It could supposedly be referred to as international capitalism, made up in the interest of Americans or other bad boys. In a way, it would be good news if this hypothesis were right. It would then be possible to envisage the means of combating the monster. Alas, this theory does not hold water. The monster is rather in each and every one of us, and it is called recklessness, ignorance and escape from reality. Combating it under these conditions will be a long and exacting endeavor to which this book can only make a modest contribution.

References

Abraham-Frois G., 1995. *Dynamique économique.* Paris, Dalloz, 654 p.

Arrow K.J., Lindt R.C., 1970. Uncertainty and the public investment decision. *In* Arrow K.J (ed.) *Essays in the theory of risk bearing.* Amsterdam, the Netherlands.

Bairoch P., 1992. *Le tiers-monde dans l'impasse.* Paris, Gallimard, Folio actuel, 672 p.

Bernoulli D., 1738. *Specimen theoriae novae de mensura sortis,* Ed. 1967. Farnborough, Gregg, 36 p.

Bertrand, J.P., Delorme H., 2007. Pratique de la régulation des marchés agricoles internationaux : le cas du blé et du soja. in: Boussard and Delorme (ed.) (2007): *La régulation des marchés agricoles* L'Harmattan, Paris.

Bouët A., 2001. *La fin de l'exception agricole.* Paris, La Découverte.

Boussard J.-M., 1975. La production agricole française: un modèle historicostatistique. Paris, Inra.

Boussard J.-M., 1996. When risk generates chaos. *Journal of economic behaviour and organisation,* 29: 433-446.

Boussard J.-M., Christensen A.K., 1996. Etude des développements possibles des économies est-européennes sous différents régimes de gestion des marchés agricoles. Miméo, Paris, Inra, 90 p.

Boussard J.-M., Gérard F., Piketty M.-G., Christensen A.K., Fallot A., Voituriez T., 2002. Modèle macroéconomique à dominante agricole par l'analyse de l'impact du changement climatique et des effets des politiques en terme d'efficacité et d'équité. Rapport final GICC, Cirad, Ministère de l'environnement.

Burton M., 1993. Some illustration of chaos in commodity models. *Journal of agricultural economics*, 44 (1): 38-50.

Cantillon R., 1755. *Essai sur la nature du commerce en général.* www.ecn.bris.ac.uk/het/cantillon.

Charman A.J.E., 2004. Malawi country study detailed findings. Document prepared for SAFP (Southern and East Africa policy assistance unit), Harare, Zimbabwe.

Chavas J.-P., Holt M.T., 1993. Market instability and non linear dynamics. *AJAE*, 75: 113-120.

Coase R.H., 1937. The nature of the firm. *Economica*, 4: 386- 405.

Cordier J., 2001. Assurance, marchés financiers et politiques publiques. *Économie rurale*, n° 266.

Costanzo S. Di, 2001. Le rôle du stockage dans la dynamique des prix des matières premières. Thèse, université Paris I, 332 p.

Daviron B., 2003. Comment interpréter les politiques agricoles? Critique de la nouvelle économie politique et contribution à l'élaboration d'un cadre d'analyse élargi. Working document, Montpellier, Cirad.

Daviron B., Voituriez T., 2003. Les paradoxes de la longévité du projet de stabilisation des marchés agricoles au XXe siècle: quelques enseignements de la pensée anglo-saxonne. *Économie et sociétés*, 37: 1579-1609.

Deaton A., Laroque G., 1992. On the behavior of commodity prices. *Review of economic studies*, 59: 1-23.

Drèze J., Sen A., 1989. *Hunger and public action.* London, Clarendon press.

Ezekiel M., 1938. The cobweb theorem. *Quarterly journal of economics*, 53: 225-280.

Folmer C., Keyser M.A., Merbis M.D., Stolwijk H.S.J., Veenandaal P.J.J., 1995. *The common agricultural policy and the Mc Sharry reform.* Amsterdam, Elsevier, 348 p.

Freund R.J., 1956. Introducing risk into a programming model. *Econometrica*, 21 (4): 253-263.

Galiani F., 1770. *Dialogue sur le commerce des bleds*, éd. 1984. Paris, Fayard, 192 p.

Gardner B.L., 1992. Changing economic perspectives in the farm problem. *Journal of economic literature*, 30: 62-101.

Gérard F., Boussard J.-M., 1994. Stabilisation des prix et offre agricole. *In* Benoit-Cattin M., Griffon M. et Guillaumont P. (édit.) Translation: *Economics of agricultural policies in developing countries*, 1995. Paris, éditions de la RFSP.

Hayami Y., Ruttan V.W., 1971. *Agricultural development: an international perspective.* Baltimore, John Hopkins press, 368 p.

Hazell P.R., Scandizzo P., 1977. Farmers' expectations, risk aversion and market equilibrium under risk. *American Journal of Agricultural Economics*, 59: 204-209.

Henk A.J. Moll, Henk A., Heerink N., 2003. Price adjustment and the cattle subsector in central West Africa. Discussion paper, Wageningen agricultural university, Department of economics, The Netherlands.

Hertel T., Tsigas M., 1997. Structure of GTAP. *In* Hertel T. (ed.) *Global trade analysis.* Cambridge, Cambridge University Press, 404 p.

Hölzer C., Precht M., 1993. Der chaotische Schweinezykus. *Agrarwirtschaft*, 42 (7): 276-283.

Hommes C., Sonnemans J., van de Velden H., 1998. Expectation formation in a cobweb economy; some one-person experiments. *In* Delli Gatti D., Gallegati M., Kirman A. (eds) *Market structure, aggregation and heterogeneity.* Heideberg, Berlin, Springer Verlag, p. 253-266.

Kajisa K., Akiyama T., 2003. The evolution of rice price policies over four decades: Thailand, Indonesia and the Philippines. Discussion paper, Tokyo foundation for advanced studies on international development.

Keynes J.M., 1936. *General Theory of Employment, Interest, and Money*, éd. 1997 London ; New York : Routledge. 490 pp.

Koning N., 2002. Agricultural negotiations in the WTO: history, backgrounds and implications. Miméo, Department of social sciences, Wageningen university.

Krugman P.R., Obstfeld M., 2003. *Économie internationale*. Brussels, De Boeck, 858 p.

Laroque G., Deaton A., 1992. On the behavior of commodities markets. *Review of economic studies*, 59: 1-23.

Leuthold R.M., Wei A., 1998. Long agricultural futures prices: Arch, long memory, or chaos processes? Mimeo. Ofor working papers, 98-03.

List F., 1849. *Das nationale system der politischen ökonomie*. Translation: *National System of Political economy*, éd. 1998. Paris, Gallimard, préface et commentaires d'Emmanuel Todd.

Lücke B., 1992. *Price stabilisation on world agricultural markets. An application to the world market for sugar.* Berlin, Springer Verlag.

Martineu O., Tissot H., 1993. Répartition géographique des aléas climatiques. Miméo, Paris, Engref.

Mocilnikar A.T., 1998. Manipulations et interventions publiques sur le marché des droits à polluer. Miméo, Paris, Commissariat général du Plan.

Moll H.A.J., Heerinck N.B.M., 1998. Price adjustments and the cattle sub-sector in Central West Africa. *Proceedings of the International*

Conference on Livestock and the Environment, Ede, Wageningen (The Netherlands) 72-87.

Morlon P., 1987. Del clima a la comercialization : un riesgo puede ocultar otro. Ejemplos sobre el Altiplano peruano. *Agricultura y sociedad*, 45 : 133-177.

Nerlove M., 1958. *The dynamics of supply*. Ames, United States, Iowa state university press.

Nerlove M., Grether D.M., Carvalho J.L., 1995. *Analysis of economic time series: a synthesis*. San Diego, United states, Academic press.

Newbery D.M.G., Stiglitz J.E., 1981. *The theory of commodity price stabilization*. Oxford, Clarendon press, 462 p.

Oi W., 1961. The desirability of price instability under perfect competition. *Econometrica*, 29: 58-64.

Olson M., 1965. *The logic of collective action*, Harvard university press.

Peter G., 2001. *Colin Clark (1905-1986), economist and agricultural economist*. Oxford, university of Oxford, Queen Elisabeth House working paper series, n° 69, 17 p.

Ricardo D., 1817. *On the principles of political economy and taxation* ed. 1822, London, Murray 95 pp.

Roll R., 1984. Orange juice and weather. *American economic review*, 74 (5): 861-880.

Schumpeter J., 1942. *Capitalism, socialism and democracy*, London, Allen and Unwind, 454 p.

Schumpeter J., 1954. *History of economic analysis*. London, Allen and Unwin, 1 260 p.

Smith A., 1776. *An Inquiry into the Nature and Causes of The Wealth of Nations*. 1993 edition: Rowman and Littlefield, Savage (Md) 599 pp.

Walras L., 1885. Théorie mathématique du prix des terres et de leur rachat par l'État. *Bulletin de la Société vaudoise des sciences naturelles*, 171 (11) : 231.

Wicksell K., 1898. *Geldzins und Güterpreise. Eine Untersuchung über die den Tauschwert des Geldes bestimmenden Ursachen*. Jena, Gustav Fischer. Translation: *Interest and prices. A study of the causes regulating the value of money*, 1936 ed. London, Macmillan, 190 p.

Williams J., 2001. Agricultural commodity markets: spot, futures, option, forward contracts and derivatives. *In* Gardner B.L. and Rausser G.C. (eds) *Handbook of agricultural economics*, vol. 1B. Amsterdam, the Netherlands, Elsevier, 564 p.

Zajdenweber D., 2001. *Économie des extrêmes*. Paris, Flammarion, 214 p.

Zanden J.L. Van, 1991. The First Green Revolution: The Growth of Production and Productivity in European Agriculture, 1870- 1914. *Economic History Review*, 44: 215-239.